PE Mechanical

HVAC and Refrigeration Practice Exam

Michael R. Lindeburg, PE

PPI2PASS.COM
A **KAPLAN** COMPANY

Register Your Book at ppi2pass.com

- Receive the latest exam news.
- Obtain exclusive exam tips and strategies.
- Receive special discounts.

Report Errors for This Book

PPI is grateful to every reader who notifies us of a possible error. Your feedback allows us to improve the quality and accuracy of our products. Report errata at **ppi2pass.com**.

Digital Book Notice

All digital content, regardless of delivery method, is protected by U.S. copyright laws. Access to digital content is limited to the original user/assignee and is non-transferable. PPI may, at its option, revoke access or pursue damages if a user violates copyright law or PPI's end-user license agreement.

PE MECHANICAL HVAC AND REFRIGERATION PRACTICE EXAM

Current release of this edition: 3

Release History

date	edition number	revision number	update
Oct 2017	1	1	New book.
Oct 2018	1	2	Minor corrections. Minor cover updates.
Jun 2019	1	3	Minor cover updates.

© 2019 Kaplan, Inc. All rights reserved.

All content is copyrighted by Kaplan, Inc. No part, either text or image, may be used for any purpose other than personal use. Reproduction, modification, storage in a retrieval system or retransmission, in any form or by any means, electronic, mechanical, or otherwise, for reasons other than personal use, without prior written permission from the publisher is strictly prohibited. For written permission, contact permissions@ppi2pass.com.

Printed in the United States of America.

PPI
1250 Fifth Avenue, Belmont, CA 94002
(650) 593-9119
ppi2pass.com

ISBN: 978-1-59126-540-5

Library of Congress Control Number: 2017954341

Table of Contents

Preface and Acknowledgments

By this time, if you are at a stage in your preparation where you are picking up *PE Mechanical HVAC and Refrigeration Practice Exam* and reading this preface (most people don't), then you probably have already read the prefaces in the *Mechanical Engineering Reference Manual* and *Practice Problems for the Mechanical Engineering PE Exam*. In that case, you will have read all the clever and witty things I wrote in those prefaces, so I won't be able to reuse them here.

This book accurately reflects the NCEES Mechanical—HVAC and Refrigeration exam specifications and multiple-choice format. The format of the PE exam presents several challenges to examinees. The breadth of the subject matter covered by the exam requires that you have a firm grasp of mechanical engineering fundamentals. These rudiments are generally covered in an undergraduate curriculum. You still need to know how to select a pump, size a shaft, and humidify a room. However, exam problems enable NCEES to focus on, target, and drill down to some very specific knowledge bases. The problems within each area of emphasis require knowledge gained only through experience—a true test of your worthiness of licensure. Often, problems testing proficiency in an area can be worded as simple definition questions. For example, if you don't recognize the application of a jockey pump, you probably aren't ready to design fire protection sprinkler systems.

Needless to say, the number of problems in this practice exam is consistent with the NCEES exam, with 80 total problems. Like the exam, this book exclusively uses U.S. customary system (USCS) units. The solutions in this practice exam are also consistent in nomenclature and style with the *Mechanical Engineering Reference Manual*. Hopefully, you have already made that book part of your exam-day arsenal.

While *PE Mechanical HVAC and Refrigeration Practice Exam* reflects the NCEES exam specifications, none of the problems in this book are actual exam problems. The problems in this book have come directly from my imagination and the imagination of my colleagues, Keith Elder, PE, and James Kamm, PhD, PE, LEED AP. Material was technically reviewed by Anthony Darmiento, PE. I am grateful for their efforts.

Editing, typesetting, illustrating, and proofreading of this book continues to follow PPI's strict style guidelines for engineering publications. Managed by Grace Wong, director of publishing services, the following top-drawer individuals have my thanks for bringing this book to life: Nancy Peterson, project manager; Cathy Schrott, production services manager; Jenny Lindeburg King, editor-in-chief; Scott Marley, copy editor; Scott Rutherford, proofreader; Darren Meiss, freelance proofreader; Robert Genevro and Ellen Nordman, typesetters; Tom Bergstrom, technical illustrator and cover designer; and Anil Acharya, calculation checker. Without their efforts, this book would read like a bunch of scribbles.

As in all of my publications, I invite your comments. If you disagree with a solution, or if you think there is a better way to do something, please let me know. You can submit errata online at the PPI website at **ppi2pass.com/errata**.

Best wishes in your exam and subsequent career.

Michael R. Lindeburg, PE

Codes Used to Prepare This Book

The documents, codes, and standards that I used to prepare this book were the most current available at the time of publication. In the absence of any other specific need, that was the best strategy.

Engineering practice is often constrained by law or contract to using codes and standards that have already been adopted or approved. However, newer codes and standards might be available. For example, the adoption of building codes by states and municipalities often lags publication of those codes by several years. By the time the 2015 codes are adopted, the 2017 codes have been released. Federal regulations are always published with future implementation dates. Contracts are signed with designs and specifications that were "best practice" at some time in the past. Nevertheless, the standards are referenced by edition, revision, or date. All of the work is governed by unambiguous standards.

All standards produced by ASME, ASHRAE, ANSI, ASTM, and similar organizations are identified by an edition, revision, or date. However, although NCEES lists "codes and standards" in its lists of Mechanical—HVAC and Refrigeration exam specifications, no editions, revisions, or dates are specified. My conclusion is that the exam is not sensitive to changes in codes, standards, regulations, or announcements in the Federal Register. This is the reason I referred to the most current documents available as I prepared this book.

Introduction

ABOUT THE PE EXAM

PE Mechanical HVAC and Refrigeration Practice Exam provides the opportunity to practice taking an eight-hour exam similar in content and format to the Principles and Practice of Engineering (PE) Mechanical—HVAC and Refrigeration exam. The exam is eight-hours divided into a morning session and an afternoon session. In the four-hour morning session, you are asked to solve 40 problems and in the four-hour afternoon session you are also asked to solve 40 problems. All in all, you will be solving 80 problems in eight hours.

All problems are multiple-choice and in U.S. customary system (USCS) units. They include a problem statement with all the required information, followed by four logical choices. Only one of the four options is correct. The problems are completely independent from each other, so an incorrect choice on one problem will not carry over to subsequent problems.

The problems for each session are drawn from either the Principles or Applications knowledge areas as specified by the National Council of Examiners for Engineering and Surveying (NCEES). The exam knowledge areas and the approximate number of problems are as follows.

Principles (32 problems)

- Basic Engineering Practice (4 problems)

- Thermodynamics (4 problems)

- Psychrometrics (8 problems)

- Heat Transfer (7 problems)

- Fluid Mechanics (4 problems)

- Energy/Mass Balance (5 problems)

Applications (48 problems)

- Heating/Cooling Loads (8 problems)

- Equipment and Components (18 problems)

- Systems and Components (18 problems)

- Supportive Knowledge (4 problems)

HOW TO USE THIS BOOK

This book is a practice exam—the main issue is not how you use it, but when you use it. It was not intended to be a diagnostic tool to guide your preparation. Rather, its value is in giving you an opportunity to bring together all of your knowledge and to practice your test-taking skills. The three most important skills are (1) selection of the right subjects to study, (2) organization of your references and other resources, and (3) time management. Take this practice exam within a few weeks of your actual exam. That's the only time that you will be able to focus on test-taking skills without the distraction of rusty recall.

Do not read the questions ahead of time, and do not look at the answers until you've finished. Prepare for the practice exam as you would prepare for the actual exam. Assemble your reference materials. Check with your state's board of engineering registration for any restrictions on what materials you can bring to the exam. (The PPI website has a listing of state boards at **ppi2pass.com/faqs/state-boards**.)

Read the practice exam instructions (which simulate the ones you'll receive from your exam proctor), set a timer for four hours, and answer the first 40 problems. After a one-hour break, turn to the last 40 problems in this book, set the timer, and complete the simulated afternoon session. Then, check your answers.

The problems in this book were written to emphasize the breadth of the HVAC and refrigeration mechanical engineering field. Some may seem easy and some hard. If you are unable to answer a problem, you should review that topic area in the *Mechanical Engineering Reference Manual for the PE Exam.*

The problems are generally similar to each other in difficulty, yet a few somewhat easier problems have been included to expose you to less frequently examined topics. On the exam, only your answer sheet will be scored. No credit will be given for calculations written in your exam booklet.

While some topics in this book may not appear on your exam, the concepts and problem style will be useful practice.

Once you've taken the full exam, check your answers. Evaluate your strengths and weaknesses, and select additional texts to supplement your weak areas (e.g., *Practice Problems for the Mechanical Engineering PE Exam*). Check the PPI website for the latest in exam preparation materials at **ppi2pass.com**.

The keys to success on the exam are to know the basics and to practice solving as many problems as possible. This book will assist you with both objectives.

Pre-Test

You can use the following incomplete table to judge your preparedness. You should be able to fill in all of the missing information. (The completed table appears on the back of this page.) If you are ready for the sample exam (and, hence, for the actual exam), you will recognize them all, get most correct, and when you see the answers to the ones you missed, you'll say, "Ahh, yes." If you have to scratch your head with too many of these, then you haven't exposed yourself to enough of the subjects that are on the exam.

description	value or formula	units
acceleration of gravity, g		in/sec^2
gravitational constant, g_c	32.2	
formula for the area of a circle		ft^2
	1545	ft-lbf/lbmol-°R
	+460	°
density of water, approximate		lbm/ft^3
density of air, approximate	0.075	
specific gas constant for air	53.35	
	1.0	Btu/lbm-°R
foot-pounds per second in a horsepower		ft-lbf/hp-sec
pressure, p, in fluid with density, ρ, in lbm/ft^3, at depth, h		lbf/ft^2
cancellation and simplification of the units (A = amps; rad = radians)	A-sec^4/sec^5-rad	
specific heat of air, constant pressure		Btu/lbm-°R
common units of entropy of steam	–	
	$bh^3/12$	cm^4
molecular weight of oxygen gas		lbm/lbmol
what you add to convert ΔT_{F} to ΔT_{R}		°
	Q/A	ft/sec
inside surface area of a hollow cylinder with length L and diameters d_i and d_o		ft^2
the value that NPSHA must be larger than		ft
	849	lbm/ft^3
what you have to multiply density in lbm/ft^3 by to get specific weight in lbf/ft^3		lbf/lbm
power dissipated by a device drawing I amps when connected to a battery of V volts		W
linear coefficient of thermal expansion for steel	6.5×10^{-6}	
formula converting degrees centigrade to degrees Celsius		
the primary SI units constituting a newton of force		N
the difference between psig and psia at sea level		psi
the volume of a mole of an ideal gas		ft^3
	1.71×10^{-9}	Btu/ft^2-hr-°R^4
universal gas constant	8314.47	
	2.31	ft/psi
shear modulus of steel		psi
conversion from rpm to rad/sec		rad-min/rev-sec
Joule's constant	778.17	

Pre-Test Answer Key

description	value or formula	units
acceleration of gravity, g	386	in/sec^2
gravitational constant, g_c	32.2	lbm-ft/lbf-sec^2
formula for the area of a circle	πr^2 or $(\pi/4)d^2$	ft^2
universal gas constant in customary U.S. units	1545	ft-lbf/lbmol-°R
what you add to $T_{°F}$ to obtain $T_{°R}$ (absolute temperature)	+460	°
density of water, approximate	62.4	lbm/ft^3
density of air, approximate	0.075	lbm/ft^3
specific gas constant for air	53.35	ft-lbf/lbm-°R
specific heat of water	1.0	Btu/lbm-°R
foot-pounds per second in a horsepower	550	ft-lbf/hp-sec
pressure, p, in fluid with density, ρ, in lbm/ft^3, at depth, h	$p = \gamma h = \rho g h / g_c$	lbf/ft^2
cancellation and simplification of the units (A = amps; rad = radians)	A-sec^4/sec^5-rad	W (watts)
specific heat of air, constant pressure	0.241	Btu/lbm-°R
common units of entropy of steam	–	Btu/lbm-°R
centroidal moment of inertia of a rectangle	$bh^3/12$	cm^4
molecular weight of oxygen gas	32	lbm/lbmol
what you add to convert $\Delta T_{°F}$ to $\Delta T_{°R}$	0	°
velocity of flow	Q/A	ft/sec
inside surface area of a hollow cylinder with length L and diameters d_i and d_o	$\pi d_i L$	ft^2
the value that NPSHA must be larger than	h_v (vapor head)	ft
density of mercury	849	lbm/ft^3
what you have to multiply density in lbm/ft^3 by to get specific weight in lbf/ft^3	g/g_c (numerically, 32.2/32.2 or 1.0)	lbf/lbm
power dissipated by a device drawing I amps when connected to a battery of V volts	IV	W
linear coefficient of thermal expansion for steel	6.5×10^{-6}	1/°F or 1/°R
formula converting degrees centigrade to degrees Celsius	°centigrade = °Celsius	The centigrade scale is obsolete.
the primary SI units constituting a newton of force	kg·m/s^2	N
the difference between psig and psia at sea level	14.7 psia (atmospheric pressure)	psi
the volume of a mole of an ideal gas	359 or 360	ft^3
Stefan-Boltzmann constant	1.71×10^{-9}	Btu/ft^2-hr-°R^4
universal gas constant	8314.47	J/kmol·K
conversion from psi to height of water	2.31	ft/psi
shear modulus of steel	11.5×10^6	psi
conversion from rpm to rad/sec	$2\pi/60$	rad-min/rev-sec
Joule's constant	778.17	ft-lbf/Btu

Practice Exam Instructions

In accordance with the rules established by your state, you may use textbooks, handbooks, bound reference materials, and any approved battery- or solar-powered, silent calculator to work this examination. However, no blank papers, writing tablets, unbound scratch paper, or loose notes are permitted. Sufficient room for scratch work is provided in the Examination Booklet.

You are not permitted to share or exchange materials with other examinees.

You will have eight hours for the examination: four hours to answer the first 40 questions, a one hour lunch break, and four hours to answer the second 40 questions. Your score will be determined by the number of questions that you answer correctly. There is a total of 80 questions. All 80 questions must be worked correctly in order to receive full credit on the exam. There are no optional questions. Each question is worth 1 point. The maximum possible score for the examination is 80 points.

Partial credit is not available. No credit will be given for methodology, assumptions, or work written in your Examination Booklet.

Record all of your answers on the Answer Sheet. No credit will be given for answers marked in the Examination Booklet. Mark your answers with a no. 2 pencil. Answers marked in pen may not be graded correctly. Marks must be dark and must completely fill the bubbles. Record only one answer per question. If you mark more than one answer, you will not receive credit for the question. If you change an answer, be sure the old bubble is erased completely; incomplete erasures may be misinterpreted as answers.

If you finish early, check your work and make sure that you have followed all instructions. After checking your answers, you may turn in your Examination Booklet and Answer Sheet and leave the examination room. Once you leave, you will not be permitted to return to work or change your answers.

When permission has been given by your proctor, break the seal on the Examination Booklet. Check that all pages are present and legible. If any part of your Examination Booklet is missing, your proctor will issue you a new Booklet.

WAIT FOR PERMISSION TO BEGIN.

Name: _____
   Last   First   Middle Initial

Examinee number: _____

Examination Booklet number: _____

Principles and Practice of Engineering Examination

Practice Examination

1. Ⓐ Ⓑ Ⓒ Ⓓ
2. Ⓐ Ⓑ Ⓒ Ⓓ
3. Ⓐ Ⓑ Ⓒ Ⓓ
4. Ⓐ Ⓑ Ⓒ Ⓓ
5. Ⓐ Ⓑ Ⓒ Ⓓ
6. Ⓐ Ⓑ Ⓒ Ⓓ
7. Ⓐ Ⓑ Ⓒ Ⓓ
8. Ⓐ Ⓑ Ⓒ Ⓓ
9. Ⓐ Ⓑ Ⓒ Ⓓ
10. Ⓐ Ⓑ Ⓒ Ⓓ
11. Ⓐ Ⓑ Ⓒ Ⓓ
12. Ⓐ Ⓑ Ⓒ Ⓓ
13. Ⓐ Ⓑ Ⓒ Ⓓ
14. Ⓐ Ⓑ Ⓒ Ⓓ
15. Ⓐ Ⓑ Ⓒ Ⓓ
16. Ⓐ Ⓑ Ⓒ Ⓓ
17. Ⓐ Ⓑ Ⓒ Ⓓ
18. Ⓐ Ⓑ Ⓒ Ⓓ
19. Ⓐ Ⓑ Ⓒ Ⓓ
20. Ⓐ Ⓑ Ⓒ Ⓓ
21. Ⓐ Ⓑ Ⓒ Ⓓ
22. Ⓐ Ⓑ Ⓒ Ⓓ
23. Ⓐ Ⓑ Ⓒ Ⓓ
24. Ⓐ Ⓑ Ⓒ Ⓓ
25. Ⓐ Ⓑ Ⓒ Ⓓ
26. Ⓐ Ⓑ Ⓒ Ⓓ
27. Ⓐ Ⓑ Ⓒ Ⓓ

28. Ⓐ Ⓑ Ⓒ Ⓓ
29. Ⓐ Ⓑ Ⓒ Ⓓ
30. Ⓐ Ⓑ Ⓒ Ⓓ
31. Ⓐ Ⓑ Ⓒ Ⓓ
32. Ⓐ Ⓑ Ⓒ Ⓓ
33. Ⓐ Ⓑ Ⓒ Ⓓ
34. Ⓐ Ⓑ Ⓒ Ⓓ
35. Ⓐ Ⓑ Ⓒ Ⓓ
36. Ⓐ Ⓑ Ⓒ Ⓓ
37. Ⓐ Ⓑ Ⓒ Ⓓ
38. Ⓐ Ⓑ Ⓒ Ⓓ
39. Ⓐ Ⓑ Ⓒ Ⓓ
40. Ⓐ Ⓑ Ⓒ Ⓓ
41. Ⓐ Ⓑ Ⓒ Ⓓ
42. Ⓐ Ⓑ Ⓒ Ⓓ
43. Ⓐ Ⓑ Ⓒ Ⓓ
44. Ⓐ Ⓑ Ⓒ Ⓓ
45. Ⓐ Ⓑ Ⓒ Ⓓ
46. Ⓐ Ⓑ Ⓒ Ⓓ
47. Ⓐ Ⓑ Ⓒ Ⓓ
48. Ⓐ Ⓑ Ⓒ Ⓓ
49. Ⓐ Ⓑ Ⓒ Ⓓ
50. Ⓐ Ⓑ Ⓒ Ⓓ
51. Ⓐ Ⓑ Ⓒ Ⓓ
52. Ⓐ Ⓑ Ⓒ Ⓓ
53. Ⓐ Ⓑ Ⓒ Ⓓ
54. Ⓐ Ⓑ Ⓒ Ⓓ

55. Ⓐ Ⓑ Ⓒ Ⓓ
56. Ⓐ Ⓑ Ⓒ Ⓓ
57. Ⓐ Ⓑ Ⓒ Ⓓ
58. Ⓐ Ⓑ Ⓒ Ⓓ
59. Ⓐ Ⓑ Ⓒ Ⓓ
60. Ⓐ Ⓑ Ⓒ Ⓓ
61. Ⓐ Ⓑ Ⓒ Ⓓ
62. Ⓐ Ⓑ Ⓒ Ⓓ
63. Ⓐ Ⓑ Ⓒ Ⓓ
64. Ⓐ Ⓑ Ⓒ Ⓓ
65. Ⓐ Ⓑ Ⓒ Ⓓ
66. Ⓐ Ⓑ Ⓒ Ⓓ
67. Ⓐ Ⓑ Ⓒ Ⓓ
68. Ⓐ Ⓑ Ⓒ Ⓓ
69. Ⓐ Ⓑ Ⓒ Ⓓ
70. Ⓐ Ⓑ Ⓒ Ⓓ
71. Ⓐ Ⓑ Ⓒ Ⓓ
72. Ⓐ Ⓑ Ⓒ Ⓓ
73. Ⓐ Ⓑ Ⓒ Ⓓ
74. Ⓐ Ⓑ Ⓒ Ⓓ
75. Ⓐ Ⓑ Ⓒ Ⓓ
76. Ⓐ Ⓑ Ⓒ Ⓓ
77. Ⓐ Ⓑ Ⓒ Ⓓ
78. Ⓐ Ⓑ Ⓒ Ⓓ
79. Ⓐ Ⓑ Ⓒ Ⓓ
80. Ⓐ Ⓑ Ⓒ Ⓓ

Practice Exam

1. The water level of a 15 psig pressurized tank is 8 ft below the water level of an open tank. A pump 15 ft below the water level of the open tank delivers water through schedule-40 steel pipe to the pressurized tank. The losses due to both pipe friction and fittings are 12 ft. The pump must have a total dynamic head that is most nearly

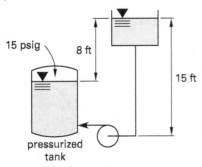

(A) 19 ft water

(B) 27 ft water

(C) 39 ft water

(D) 54 ft water

2. When running at full load, a chiller has the capacity to cool 640 gpm of water from 55°F to 43°F. If the rated coefficient of performance (COP) is 4.4, the total compressor heat that must be rejected to the cooling tower is most nearly

(A) 21 tons

(B) 73 tons

(C) 320 tons

(D) 390 tons

3. The kinematic viscosity of an SAE 10W-30 engine oil with a specific gravity of 0.88 is reported as 110 centistokes (cSt) at 37°C. Expressed in Saybolt universal seconds (SUS), the viscosity is most nearly

(A) 0.90 SUS

(B) 49 SUS

(C) 510 SUS

(D) 32,000 SUS

4. An environmental criterion for a museum requires 8000 ft^3/min supply air to be delivered to a gallery space at 60°F db and 51°F dew point. If the condition of the air leaving the cooling coil is 53°F db and 52°F wb, the sensible reheat load is most nearly

(A) 22,000 Btu/hr

(B) 60,000 Btu/hr

(C) 78,000 Btu/hr

(D) 84,000 Btu/hr

5. A fluid flows through a tube at a rate of 3×10^{-5} ft^3/sec. The tube is 3.4 ft long and has an internal diameter of 0.018 in. The flow is viscous with a viscosity of 26.37×10^{-6} lbf-sec/ft^2, incompressible, steady, and laminar. The predicted pressure drop across the length of the tube is most nearly

(A) 11,000 lbf/ft^2

(B) 15,000 lbf/ft^2

(C) 22,000 lbf/ft^2

(D) 26,000 lbf/ft^2

6. Investors are evaluating an office building that has a purchase price of \$525,000. The investors expect to sell the building for \$700,000 after 7 years. During the time that the building is held, the building will be depreciated using straight line depreciation, an estimated life of 15 years, and an (assumed) salvage value of zero. (The full purchase price will be depreciated.) The effective tax rate of the investors is 40%.

During ownership, the estimated expenses will be \$25,000 per year, which will be offset by rental income of \$45,000 for the first 3 years and \$75,000 for the next 4 years.

Neglect the benefit of a favorable capital gains tax rate on the gain above \$525,000. The after-tax rate of return on this investment will be most nearly

(A) 2.1%

(B) 7.1%

(C) 7.9%

(D) 12%

7. A 20 in diameter suction line carries water at 7000 gpm. The water enters a pump at 100 psi and is pumped up a 30 ft high incline through a 12 in diameter discharge line. If the pressure at the top of the incline is 500 psi and the head loss through the 30 ft section is 10 ft, the horsepower that is supplied by the pump is most nearly

(A)   140 hp

(B)   750 hp

(C)   1200 hp

(D)   1700 hp

8. Windows to be installed in a building are specified with glass having an R-value of 1.5 hr-ft^2-°F/Btu. Outdoor temperatures as low as -40°F are experienced in the winter, and the indoor space is to be heated to 70°F. Assuming indoor and outdoor window film coefficients of 1.46 Btu/hr-ft^2-°F and 6.0 Btu/hr-ft^2-°F, respectively, the maximum relative humidity that can be maintained without experiencing condensation on the inside of the glass is most nearly

(A)   31%

(B)   34%

(C)   37%

(D)   42%

9. Air with a mass of 3 lbm is expanded polytropically behind a piston in a cylinder from an initial condition of 120 psia and 100°F to a final condition of 40 psia. Using a polytropic exponent of 1.6, the work done by the closed system per unit mass is most nearly

(A) -1.7×10^4 ft-lbf/lbm

(B) -5.6×10^3 ft-lbf/lbm

(C) 5.6×10^3 ft-lbf/lbm

(D) 1.7×10^4 ft-lbf/lbm

10. A heat exchanger pump receives water at 140°F and returns it at 95°F. The water is received and returned through nominal 3 in copper tubing (with an outside diameter of 3.5 in, inside diameter of 3.062 in, cross-sectional area of 7.370 in^2, and electrical resistivity of 10 Ω-cmil/ft). The pump is driven by a one-phase, 110 V AC, 5 hp motor that draws an average (RMS) current of 35 A. The engineer decides to place the copper tubing inside PVC pipe for electrical insulation, and use the copper tubing to carry the electrical current that will power the pump motor. The maximum distance that the pump

and motor can be located from a 110 V source so that the voltage drop does not exceed 0.6 V is most nearly

(A)   870 ft

(B)   2500 ft

(C)   4900 ft

(D)   9000 ft

11. A 2 hp motor is used to stir a tank containing 60 lbm of water for 15 min. Assuming the process occurs at constant volume, the maximum possible rise in temperature is most nearly

(A)   1.4°F

(B)   21°F

(C)   58°F

(D)   130°F

12. The rectangular wall of a furnace is made from 3 in fire-clay brick ($k = 0.58$ Btu-ft/hr-ft^2-°F) surrounded by 0.25 in of steel on the outside. 0.25 in diameter mild steel bolts connect the steel to the brick. The furnace is surrounded by 70°F air with a convection coefficient of 1.65 Btu/hr-ft^2-°F, while the inner surface of the brick is held constant at 1000°F. Disregard conductance through the bolts. The outside surface temperature of the steel is most nearly

(A)   160°F

(B)   470°F

(C)   610°F

(D)   760°F

13. A parallel-flow tube-and-shell heat exchanger is designed using 1 in OD tubing. 40,000 lbm/hr of water at 55°F are used to cool 45,000 lbm/hr of a 95% ethyl alcohol solution ($c_p = 0.9$ Btu/lbm-°F) from 160°F to 110°F. If the overall coefficient of heat transfer based on the outer tube area is 75 Btu/hr-ft^2-°F, then the heat-transfer surface area of the heat exchanger is most nearly

(A)   520 ft^2

(B)   850 ft^2

(C)   920 ft^2

(D)   1100 ft^2

14. A steam turbine operates as a component of a Rankine cycle. Steam is supplied to the turbine at 1000 psia and 800°F. The turbine exhausts at 4 psia. The expansion is not reversible, and the exhaust is vapor at 100%

quality. The thermal efficiency of the turbine is most nearly

(A) 59%

(B) 76%

(C) 85%

(D) 100%

15. The gravimetric air/fuel ratio of methane burned with 125% theoretical air is most nearly

(A) 13:1

(B) 17:1

(C) 22:1

(D) 35:1

16. A solid copper sphere 200 mm in diameter is at a temperature of 450°C. It is dropped into a large tank of 75°C oil. If the average convective heat transfer coefficient is 880 W/m²·K and the oil is stirred uniformly at all times, most nearly how long after submersion does the sphere reach a temperature of 200°C?

(A) 5.3 s

(B) 160 s

(C) 910 s

(D) 4700 s

17. A refrigeration system using R-134a refrigerant operates at 33 psia on the low-pressure side and 100 psia on the high-pressure side. The 255 ton system delivers refrigerant vapor with 20°F superheat to the compressor. If the refrigerant on the high-pressure side is cooled to saturated liquid before expansion, the required refrigerant mass flow rate is most nearly

(A) 43,000 lbm/hr

(B) 48,000 lbm/hr

(C) 51,000 lbm/hr

(D) 55,000 lbm/hr

18. 7000 ft³/min of recirculated air from a room conditioned to 75°F and 50% relative humidity are mixed with 2300 ft³/min of outdoor air at 90°F db and 75°F wb prior to passing through a cooling coil and fan for

distribution back to the room. The humidity ratio of the air entering the coil is most nearly

(A) 0.0093 lbm moisture/lbm dry air

(B) 0.011 lbm moisture/lbm dry air

(C) 0.014 lbm moisture/lbm dry air

(D) 0.015 lbm moisture/lbm dry air

19. A centrifugal pump with a net positive suction head requirement (NPSHR) of 12 ft water is required to deliver 80°F water from a lake to a storage reservoir. The lake and the reservoir are located near sea level. The reservoir surface is 17 ft above the lake surface. The pump itself is located near the surface of the storage reservoir. Friction and fitting losses are estimated to be 3.0 ft water. The net positive suction head available (NPSHA) for the pump is most nearly

(A) 13 ft water

(B) 14 ft water

(C) 16 ft water

(D) 20 ft water

20. A centrifugal pump is driven at 1300 rpm by a 10 hp motor and delivers 250 gpm of 85°F water against 75 ft water head. Assuming that the initial pump efficiency of 65% does not vary appreciably, the maximum flow the pump can deliver is most nearly

(A) 280 gpm

(B) 470 gpm

(C) 520 gpm

(D) 650 gpm

21. An air conditioning system is being designed for a 40,000 ft² casino which will operate 24 hr/day. The space is completely inside a larger building and has no windows or walls exposed to direct sun on the exterior of the casino. The designer anticipates that the lighting system will require approximately 3.75 W/ft². The gaming equipment, cameras, and miscellaneous equipment require an additional 80 kW. The steam tables for a large all-day buffet contribute an additional 50,000 Btu/hr of latent load. The maximum density for patrons is estimated to be 120 people/1000 ft². It is desired to maintain space conditions at 73°F and 45% relative humidity. The total space cooling load due to internal heat gain is most nearly

(A) 890,000 Btu/hr

(B) 1,900,000 Btu/hr

(C) 2,400,000 Btu/hr

(D) 3,000,000 Btu/hr

22. An office space measures 12 ft × 12 ft and has a 9 ft ceiling. The cooling load has been determined to be 3000 Btu/hr. A single ceiling diffuser will be installed in the center of the room ceiling. To achieve good occupant comfort and a maximum air distribution performance index (ADPI), a diffuser should be selected with a throw of most nearly

(A) 5.0 ft

(B) 6.0 ft

(C) 10 ft

(D) 12 ft

23. After a driver parked and exited her automobile, the air-conditioned interior was 70°F at 40% relative humidity. After the car sat in the driveway overnight, the interior temperature dropped to 50°F, bringing the interior's relative humidity to most nearly

(A) 40%

(B) 60%

(C) 80%

(D) 100%

24. A fan supplies 4500 ft^3/min of air through a 240 ft long rectangular duct. The duct dimensions are 18 in × 24 in. If the duct has a friction factor of 0.016 and a roughness of 0.0003 ft, the total static pressure drop due to friction is most nearly

(A) 0.2 in water

(B) 0.3 in water

(C) 0.5 in water

(D) 0.6 in water

25. The water pipes in a vented crawl space are uninsulated. The space shares 320 ft^2 of exterior wall with the outdoors and 645 ft^2 of floor with a space above that is heated to 72°F. The overall heat transfer coefficients (U-factors) for the crawl space wall and floor above are

$$U_{\text{floor}} = 0.05 \text{ Btu/hr-ft}^2\text{-}°F$$
$$U_{\text{wall}} = 0.15 \text{ Btu/hr-ft}^2\text{-}°F$$

Outdoor air at 900 ft^3/hr infiltrates the crawl space through the vents. The pipes will be safe from freezing down to an outdoor temperature of most nearly

(A) −10°F

(B) 2°F

(C) 10°F

(D) 30°F

26. Water enters an evaporative cooling tower as a saturated liquid at 180°F. 8% of the water evaporates as saturated vapor. After this process, the temperature of the remaining water is most nearly

(A) 81°F

(B) 86°F

(C) 94°F

(D) 100°F

27. The water surface in a well is 90 ft below a house. A storage tank with its water surface 50 ft above the house will provide gravity water flow to the house. A submersible pump is capable of delivering 6 gpm from the well to the tank. The system is powered directly by a solar-electric array that has an overall efficiency of 12% and regularly intercepts solar radiation of a magnitude of at least 28 W/ft^2 on a cloudless day. The pump efficiency is 60%, and piping has been oversized to such an extent that friction and fitting losses are negligible. The minimum required area for the solar array operating on a cloudless day is most nearly

(A) 6 ft^2

(B) 20 ft^2

(C) 50 ft^2

(D) 80 ft^2

28. To reduce the load on a chiller plant, an air washer recirculating 57°F water is used to evaporatively pre-cool 20,000 ft^3/min outdoor air. The outdoor air is introduced at 92°F db and 57°F wb. If the saturation efficiency of the process is 84%, the cooling requirement reduction is most nearly

(A) 45 tons

(B) 49 tons

(C) 53 tons

(D) 63 tons

29. An electric unit heater is needed to heat a room on the second floor of a three-story building. The room has no windows. The 800 ft^2 external wall is made up of 2 in of polystyrene rigid insulation sandwiched between 8 in brick and ⅝ in gypsum board. The outdoor winter design temperature is 10°F, and the room is to be maintained at 74°F. Using a 25% safety factor, the required heating capacity for the unit heater is most nearly

(A) 1.5 kW

(B) 1.9 kW

(C) 2.4 kW

(D) 2.8 kW

30. 80°F air is introduced to a 3 ft × 4 ft (face-area dimensions) cooling coil at a velocity of 450 ft/min and is cooled to 56°F. The coil cooling process has a grand sensible heat ratio (GSHR) of 0.70. The total coil load is most nearly

(A) 120,000 Btu/hr

(B) 140,000 Btu/hr

(C) 180,000 Btu/hr

(D) 200,000 Btu/hr

31. In a certain system, 2530 gpm of 83°F condenser water are supplied to an 840 ton chiller for heat rejection and are returned to the cooling tower at 92°F. The coefficient of performance (COP) of the chiller is most nearly

(A) 4

(B) 5

(C) 6

(D) 8

32. Which of the following centrifugal fan impeller designs typically exhibits the highest efficiency?

(A) backward curved

(B) forward curved

(C) backward inclined

(D) airfoil

33. A 6 ft³ gas bottle holds 5 lbm of compressed nitrogen gas at 100°F. Gas is released until the bottle pressure reaches 150 lbf/in². What is most nearly the amount of gas released?

(A) 0.030 lbm

(B) 0.80 lbm

(C) 1.7 lbm

(D) 4.2 lbm

34. A store in a shopping mall is to be maintained at 75°F and 45% relative humidity with supply air of 55°F and 30% relative humidity. The space cooling load is 73,000 Btu/hr sensible and 26,000 Btu/hr latent at outdoor design conditions of 94°F db and 72°F wb. The ventilation requirement is 850 ft³/min. The coil load due to the ventilation air is most nearly

(A) 30,000 Btu/hr

(B) 60,000 Btu/hr

(C) 80,000 Btu/hr

(D) 100,000 Btu/hr

35. The reheat coil in a variable air volume (VAV) terminal box is being replaced. The maximum airflow capacity of the box is 2400 ft³/min. A minimum stop setting of 30% (of the maximum flow) has been established to maintain the required ventilation when cooling loads are at a minimum. The supply air temperature for the building system is a function of the outside air temperature, according to the reset graph shown. During the winter, the outdoor design temperature of 10°F and the indoor space temperature of 72°F result in a space heat loss of 45,000 Btu/hr. The minimum capacity of the reheat coil is most nearly

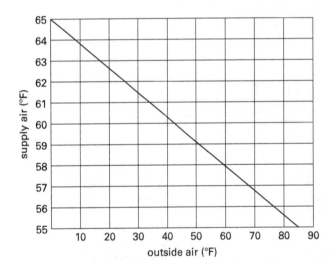

(A) 45,000 Btu/hr

(B) 47,000 Btu/hr

(C) 51,000 Btu/hr

(D) 66,000 Btu/hr

36. In a particular process, valves at each end of an 80 ft section of 2 in schedule-40 pipe are closed instantaneously, and the saturated steam (at atmospheric pressure) in the pipe begins to cool. The average heat transfer over the cooling period is 6200 Btu/hr. There is no heat loss through the valves at the ends of the pipe.

The time it takes for the steam to cool to the ambient temperature of 60°F is most nearly

(A) 0.01 min

(B) 1 min

(C) 4 min

(D) 10 min

37. A simple Rankine cycle operates between super-heated steam entering a turbine at 1200°F and 700 psia, and entering a pump at 2 psia. The cycle's maximum possible efficiency is

(A) 27%

(B) 31%

(C) 39%

(D) 43%

38. A single-stage chiller circulates 117,000 lbm/hr of R-22 refrigerant and operates with a 90°F condensing temperature and 10°F evaporating temperature. Saturated refrigerant vapor enters the compressor with no superheat, and saturated liquid refrigerant leaves the condenser with no subcooling. Heat is rejected to condenser water that enters the condenser at 85°F and leaves at 95°F. The rated coefficient of performance (COP) is 5.5 under these conditions. The condenser water flow required for heat rejection is most nearly

(A) 560 gpm

(B) 1600 gpm

(C) 1900 gpm

(D) 2200 gpm

39. An expansion tank is being provided for a steel-pipe chilled water distribution system. The system has a volume of 1500 gal and must be able to operate between temperature extremes of 50°F and 105°F. The minimum and maximum tank pressures are 10 psig and 23 psig, respectively. The steel pipe has a thermal linear expansion coefficient of 6.5×10^{-6} in/in-°F. If the tank is to be a closed type, with air/water interface, the minimum tank size is most nearly

(A) 20 gal

(B) 30 gal

(C) 50 gal

(D) 70 gal

40. Recirculated air at 7000 ft³/min, pulled from a room conditioned to 75°F and 50% relative humidity, is mixed with 2300 ft³/min of outdoor air at 90°F db and 75°F wb prior to passing through a cooling coil and a

centrifugal fan for distribution back to the room. The coil has a total cooling capacity of 300,000 Btu/hr and a sensible cooling capacity of 235,000 Btu/hr. The fan power at the shaft is 4.2 BHP. The temperature of the air leaving the coil is most nearly

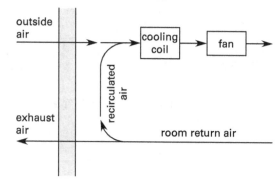

(A) 49°F

(B) 55°F

(C) 64°F

(D) 90°F

41. In the psychrometric process diagram shown, the process identified as 3-4 is best described as

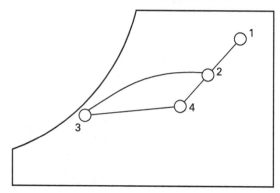

(A) evaporative cooling

(B) sensible and latent cooling

(C) dehumidification

(D) sensible and latent heating

42. The chiller system in a 1600 ft² (floor area) refrigeration machinery room contains 150 lbm of refrigerant. The room has 1.5 ft² of equivalent leakage area to the outside and an estimated heat gain of 15,000 Btu/hr. The design outside air temperature is 90°F. According to the *Uniform Mechanical Code*, refrigeration machinery rooms must be provided with dedicated mechanical exhaust with the capacity to achieve all of the following.

I. continuous maintenance of the room at 0.05 in water negative relative to adjacent spaces, Δp, with equivalent leakage area, A_e, in ft^2, as calculated by

$$\dot{V}_1 = 2610 A_e \sqrt{\Delta p}$$

II. continuous airflow of 0.5 ft^3/min per gross ft^2, A_{gf}, within the room as calculated by

$$\dot{V}_2 = 0.5 A_{gf}$$

III. temperature rise, ΔT, within the refrigeration machinery room due to heat dissipation, $\dot{q}$, in Btu/hr to a maximum of 104°F, limited by

$$\dot{V}_3 = \frac{\sum \dot{q}}{1.08 \Delta T}$$

IV. emergency purge of escaping refrigerant mass G, in lbm, as calculated by

$$\dot{V}_4 = 100 \sqrt{G}$$

The dedicated mechanical exhaust required for the room is most nearly

(A) 800 ft^3/min

(B) 900 ft^3/min

(C) 1000 ft^3/min

(D) 1200 ft^3/min

43. A water-cooled condenser heats cooling water from 65°F to 95°F. 2200 lbm/hr of saturated water vapor enters the condenser at 4 psia and exits as saturated liquid. The mass flux of the exiting cooling water is most nearly

(A) 34,000 lbm/hr

(B) 66,000 lbm/hr

(C) 74,000 lbm/hr

(D) 83,000 lbm/hr

44. Published throw information for supply air grilles and ceiling diffusers is based on the tendency of air to adhere to the ceiling under the right conditions. This tendency is called the

(A) Coanda effect

(B) Bernoulli effect

(C) Darcy effect

(D) Colebrook effect

45. A sun room, which has an exposed 6 in concrete floor slab measuring 20 ft × 30 ft, has been heated passively by solar radiation to an average temperature of 82°F by nightfall. The room thermostat setting is kept at 60°F. Assuming a floor slab convective heat transfer coefficient of 1.4 Btu/hr-ft^2-°F, the average temperature change of the slab after 1 hr of cooling is most nearly

(A) 1°F

(B) 2°F

(C) 4°F

(D) 5°F

46. A 22-unit apartment complex is to be served by a central hot water system. The units will each contain two lavatories, one bathtub, one shower, one kitchen sink, and one dishwasher. The American Society of Heating, Refrigerating and Air-Conditioning Engineers (ASHRAE) publishes the hot-water demand (gallons per hour per fixture) for these fixtures as

fixture	demand (gal/hr)
lavatory	2
bathtub	20
shower	30
kitchen sink	10
dishwasher	15

ASHRAE also recommends the use of a 0.30 demand factor and a 1.25 storage factor. The appropriate size hot-water storage tank, based upon meeting a probable 1 hr demand, is most nearly

(A) 520 gal

(B) 650 gal

(C) 1700 gal

(D) 2200 gal

47. An east-facing vertical window at a latitude of 40 degrees north has an area of 12 ft^2. The shading coefficient for the window is 0.87. The overall heat transfer coefficient is 1.2 Btu/hr-ft^2-°F. The table shown gives solar heat gain factors for 40 degrees north.

| | solar time a.m. | \multicolumn{8}{c}{solar heat gain factor (Btu/hr-ft^2)} | solar time p.m. |
		N	NE	E	SE	S	SW	W	NW	
May	0500	0	2	2	0	0	0	0	0	1900
	0600	35	127	140	70	11	11	11	11	1800
	0700	27	164	208	130	21	20	20	20	1700
	0800	26	148	218	163	30	26	26	26	1600
	0900	30	104	198	176	52	29	31	31	1500
	1000	35	53	149	164	82	34	35	35	1400
	1100	35	39	80	131	104	42	35	35	1300
	1200	38	36	41	83	112	81	39	39	1200
half day total		214	665	1023	880	357	199	175	174	
June	0500	9	20	21	5	2	2	2	2	1900
	0600	47	144	150	71	12	12	12	12	1800
	0700	36	171	206	121	21	20	20	22	1700
	0800	31	155	215	153	30	28	28	28	1600
	0900	32	113	191	160	44	33	31	31	1500
	1000	34	64	146	149	70	35	35	36	1400
	1100	37	41	80	115	89	40	39	37	1300
	1200	37	37	42	73	96	73	40	37	1200
half day total		252	735	1037	819	314	203	187	186	

On a day in May at solar time 0800, the indoor temperature is 75°F and the outdoor temperature is 42°F. The total instantaneous heat gain for this window is most nearly

(A) 480 Btu/hr

(B) 1800 Btu/hr

(C) 2300 Btu/hr

(D) 2800 Btu/hr

48. The specific heat for air at constant pressure is 0.240 Btu/lbm-°R. The specific heat for air at constant volume is 0.171 Btu/lbm-°R. The work needed to compress 10 lbm of air isentropically from atmospheric pressure and a temperature of 55°F to a pressure of 720 psia is most nearly

(A) 590 Btu

(B) 820 Btu

(C) 1600 Btu

(D) 1800 Btu

49. A building is located in a part of the country where the annual heating season is 23 weeks. The building is occupied from 8:00 a.m. until 6:00 p.m., Monday

through Friday. The building owner pays $0.30 per therm for gas heating (which includes the costs of fans, pumps, etc.), and the gas furnace efficiency is 77%. The building has the following characteristics.

internal volume	600,000 ft^3
inside temperature	70°F
ventilation rate	
occupied	1 air change per hour
unoccupied	$\frac{1}{2}$ air change per hour
walls	
area	10,000 ft^2
U-value	0.15 Btu/hr-ft^2-°F
windows	
area	2500 ft^2
U-value	1.10 Btu/hr-ft^2-°F
roof	
area	25,000 ft^2
U-value	0.06 Btu/hr-ft^2-°F
floor	
slab on grade	
exposed edge	720 linear ft
B-value	1.6 Btu/hr-ft-°F

In the past, the building owner kept the interior temperature fixed at 70°F. What will be the approximate annual savings if the building owner installs an automatic setback thermostat to reduce the interior temperature from 70°F to 58°F during unoccupied time?

(A) $930

(B) $1200

(C) $1600

(D) $1700

50. The employees of an insurance company insist that their office is too cold. When the air temperature was measured to be 72°F, they responded, "it feels colder than that in here." The single-pane window has been measured to have a surface temperature of 35°F, compared to the walls, floor, and ceiling which average 65°F. The calculated angle factor between a typical workstation and the window is 0.04. Assuming 50 ft/min air velocity, the operative temperature at the typical workstation is most nearly

(A) 64°F

(B) 68°F

(C) 69°F

(D) 70°F

51. A property owner has installed an air-cooled chiller on the side of his building, as shown. The neighboring property owner has complained that the unit is too noisy, particularly at night. The manufacturer's data states that the unit has a sound power rating of 80 dB. The local municipal code requires that the sound pressure level be no greater than 52 dB at the property line. The sound pressure level at the property line is most likely

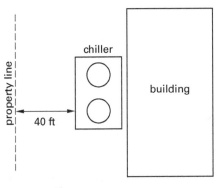

plan view

(A) 48 dB

(B) 51 dB

(C) 53 dB

(D) 57 dB

52. A fire sprinkler system uses 1 in nominal diameter schedule-40 black steel pipe. The sprinklers are located 10 ft apart. The minimum pressure at any sprinkler is 10 psig. All sprinklers have standard $\frac{1}{2}$ in orifices with discharge coefficients of 0.75. In a particular event, the last three sprinklers on a branch line are fully open simultaneously. Disregarding velocity pressure, the second branch sprinkler from the end will discharge most nearly

(A) 11 gpm

(B) 15 gpm

(C) 19 gpm

(D) 23 gpm

53. Water flows at 5 ft/sec and 60°F through 2000 ft of new, uncoated cast-iron pipe with an inner diameter of 6 in. Most nearly, the head loss is

(A) 10 ft

(B) 16 ft

(C) 24 ft

(D) 37 ft

54. During a routine inspection, a plant engineer discovers a section of bare overhead steam pipe. Upon checking the plant's maintenance records, the engineer learns that a leaking steam trap had recently been repaired, and the saturated insulation had been removed from the pipe but never replaced.

The properties of the pipe are as follows.

length of bare pipe section	120 ft
pipe material	1% carbon steel
pipe size	1.5 in BWG 16 gage
pipe mounting	ceiling pipe hangers
thermal conductivity	29 Btu/hr-ft-°F

Saturated steam at atmospheric pressure flows through the pipe at a rate high enough to prevent substantial condensation. The average inside heat transfer coefficient is 1500 Btu/hr-ft²-°F. The average outside heat transfer coefficient of the bare pipe in still air is 2.0 Btu/hr-ft²-°F. The air in the plant is at 60°F and 14.7 psia and is normally still. The pipe temperature is too low to consider the effects of radiation. The rate of heat loss from the bare pipe is most nearly

(A) 14,000 Btu/hr

(B) 18,000 Btu/hr

(C) 19,000 Btu/hr

(D) 21,000 Btu/hr

55. A home owner plans to have a contractor install a standard furnace with a rated annual fuel utilization efficiency (AFUE) of 78%. It is estimated that the furnace will consume 1500 therms of natural gas per year. The contractor has offered to install a high-efficiency condensing furnace with an AFUE of 92% for an additional $800. Assuming a constant cost of natural gas of $0.85 per therm, the simple payback period of this additional investment will be most nearly

(A) 2 yr

(B) 4 yr

(C) 5 yr

(D) 12 yr

56. The planned building envelope structure for a project has an overall coefficient of heat transfer (U-value) of 0.9 Btu/ft²-°F per hour. The price of natural gas is $6/1,000,000 Btu. The annual heating degree days for the project's location are 3500, based on a six-month heating season. The building is occupied 24 hr/day. It is proposed to add 1 in insulation board to the building envelope at a cost of $0.75/ft², which would reduce the

U-value to 0.6 Btu/ft^2-°F per hour. Most nearly, the payback period for this option is

(A) 1 yr

(B) 3 yr

(C) 4 yr

(D) 5 yr

57. A chamber is 30 ft long, 18 ft wide, and 10 ft high. The chamber contains air at 80°F, −7 psig (7.7 psia), and 40% relative humidity. Most nearly, the partial pressure of the water vapor in the chamber is

(A) 0.20 psia

(B) 0.55 psia

(C) 1.1 psia

(D) 2.1 psia

58. A refrigerant at −5°F passes through a thin cylindrical tube with a diameter of ⅝ in. Ambient air passing over the pipe is cooled such that the heat extraction rate per foot of pipe is 25 W/ft. Under steady-state conditions, the convection heat transfer coefficient is 5 W/ft^2-°F. Conductive and radiant heat transfer are negligible. Most nearly, the temperature of the ambient air is

(A) 26°F

(B) 31°F

(C) 36°F

(D) 41°F

59. Dry air flows at 30 lbm/sec into a long, insulated channel that contains a pool of water. Liquid water flowing in at point 2 is introduced as make-up water to match the evaporation rate. The dry air enters at point 1 at 80°F and atmospheric pressure, and it exits at point 3 saturated at 110°F. Most nearly, the rate of make-up water is

(A) 7.6 gpm

(B) 9.1 gpm

(C) 11 gpm

(D) 13 gpm

60. Water at 65°F flows through the pipe system shown at 48 gpm. The system uses schedule-40 pipe with a nominal diameter of ¾ in. The loss coefficients for

the 90° elbows and the 45° elbows are 0.9 and 0.42, respectively.

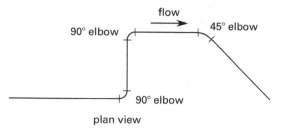

plan view

Most nearly, the sum of the minor losses for the system is

(A) 29 ft

(B) 33 ft

(C) 38 ft

(D) 44 ft

61. An air conditioning system takes in outdoor air at 40°F and 20% relative humidity at 6000 cfm. The air is sensibly heated to 70°F, followed by humidification with steam to a final condition of 90°F dry-bulb, 71°F wet-bulb. All processes are at atmospheric pressure. Most nearly, the rate of hot steam needed is

(A) 0.62 gpm

(B) 1.1 gpm

(C) 2.0 gpm

(D) 3.6 gpm

62. The pumping curve shown is for a variable speed 3450 rpm water pump. The system (or operating) curve for an application that will use this pump is superimposed on the pumping curve. At maximum load, 500 gpm at 85 ft of head is needed.

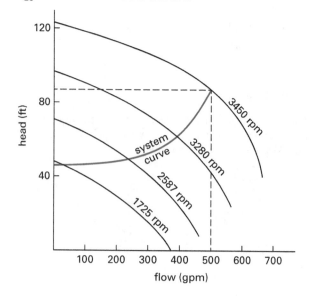

During an off-peak condition, only one-half the design flow is needed. Most nearly, the pump speed should then be reduced to

(A) 1400 rpm

(B) 1700 rpm

(C) 2600 rpm

(D) 3300 rpm

63. In which of these buildings is thermal energy storage LEAST effective?

(A) large church

(B) large refrigerated warehouse

(C) public arena

(D) enclosed retail mall

64. In a compressor filled with an ideal gas, the pressure and temperature of the suction gas are 92 psia and 25°F. The discharge pressure and temperature are 260 psia and 145°F. Most nearly, the polytropic index of this compression process is

(A) 1.1

(B) 1.2

(C) 1.3

(D) 1.4

65. Two chilled water systems are being evaluated. The first system has an energy use of 0.8 kW/ton, and the second has an energy use of 0.7 kW/ton. The installed costs of the first and second systems are $300/ton and $350/ton, respectively. The price of electricity is $0.10/kW-hr. The equivalent full load operating time is 2000 hr/yr. Most nearly, the payback period for the alternative with the higher cost efficiency is

(A) 2.5 yr

(B) 5.0 yr

(C) 7.5 yr

(D) 10 yr

66. A heat recovery wheel is placed in a system between exhaust air and makeup air. The wheel is 40% efficient and can recover only sensible heat. Air is exhausted at 75°F dry-bulb and 50% relative humidity, and makeup air enters at 35°F dry-bulb and 20% relative humidity. The rates of the exhaust and make-up

airstreams are equal. Most nearly, the amount of heat that can be recovered is

(A) 3 Btu/lbm

(B) 4 Btu/lbm

(C) 5 Btu/lbm

(D) 6 Btu/lbm

67. What is the purpose of the three-way valve shown?

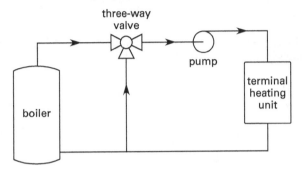

(A) modulate flow volume through the boiler

(B) modulate flow volume through the pump

(C) mix the heating water to the desired temperature

(D) maintain the constant pressure drop needed by the pump

68. In a commercial refrigeration system with a thermostatic expansion valve, which device's ONLY function is to store refrigerant?

(A) accumulator

(B) receiver

(C) expansion tank

(D) condenser

69. The most important factor in controlling *Legionella* is good management of a building's

(A) air filtration system

(B) water system

(C) refrigeration system

(D) control system

70. A round duct industrial ventilation system is shown. The static pressure at the branch is 0.10 in wg, and the exit pressure at each diffuser is 0.03 in wg. The straight branch is 20 ft long, and the take-off branch is

30 ft long. The airflow rate needed in each branch is 1000 ft³/min. Dynamic losses are negligible.

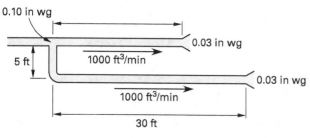

(not to scale)

Most nearly, the minimum diameters needed for the straight branch and the take-off branch, respectively, are

(A) 9 in and 12 in

(B) 9 in and 14 in

(C) 11 in and 12 in

(D) 11 in and 14 in

71. Refrigerant HFC-134a flows in a refrigeration system at 10 lbm/min. The refrigerant enters the evaporator at 20 psig (or 34.7 psia) with 15% quality, and it leaves at 15 psig (or 29.7 psia) saturated. Most nearly, the refrigerating capacity of this system is

(A) 4 tons

(B) 5 tons

(C) 6 tons

(D) 7 tons

72. Which air waste heat recovery system has the potential to recover both latent and sensible heat?

(A) run-around

(B) heat pump

(C) heat recovery wheel

(D) heat pipe

73. A 40 ton, direct expansion air conditioning system has an average electrical load of 55 kW. Most nearly, the energy efficiency ratio rating is

(A) 8.7

(B) 9.7

(C) 11

(D) 12

74. An ideal, vapor compression refrigeration cycle using refrigerant HFC-134a operates with a 90°F saturation condensing temperature and a 40°F saturation evaporator temperature. There is 10°F subcooling in the condenser and 20°F superheat at the end of the evaporator. Most nearly, the ideal horsepower needed for the compressor to achieve 3 tons of cooling is

(A) 0.5 hp

(B) 1.5 hp

(C) 2.5 hp

(D) 3.5 hp

75. Which of the following is NOT an acceptable ASHRAE duct design method?

(A) number of transfer units (NTU) method

(B) static regain method

(C) constant volume method

(D) equal friction method

76. A building's dedicated outdoor air system (DOAS) is designed to handle the entire latent load requirement for both the outside air and the building. For winter heating, the supply air condition for the building is 20,000 cfm at 105°F dry-bulb, 25% relative humidity. The inside design condition is 75°F dry-bulb, 45% relative humidity. The outside air requirement is 4000 cfm with winter design conditions of 35°F dry-bulb and 20% relative humidity. The dry-bulb temperature of the supply air for the DOAS is 75°F. Most nearly, the needed relative humidity of the supply air for the DOAS is

(A) 55%

(B) 65%

(C) 75%

(D) 85%

77. A building's HVAC system includes an air-side economizer. Under what conditions should the economizer be turned on?

(A) temperature of outside air lower than supply air

(B) temperature of outside air lower than room design condition

(C) enthalpy of outside air less than supply air enthalpy

(D) enthalpy of outside air less than room design condition

78. Which ASHRAE standard discusses methods of energy conservation?

- (A) ASHRAE 15
- (B) ASHRAE 55
- (C) ASHRAE 62
- (D) ASHRAE 90

79. The sound level at 10 ft from the source of a noise is 100 dB. Most nearly, the sound level at 80 ft from the source is

- (A) 76 dB
- (B) 82 dB
- (C) 88 dB
- (D) 94 dB

80. The ventilation system for a new 5000 ft^2 public use auditorium seating area should provide treated outdoor air of acceptable quality at a rate of most nearly

- (A) 3000 ft^3/min
- (B) 4000 ft^3/min
- (C) 6000 ft^3/min
- (D) 10,000 ft^3/min

Answer Key

1. Ⓐ Ⓑ ● Ⓓ	28. Ⓐ Ⓑ ● Ⓓ	55. Ⓐ ● Ⓒ Ⓓ
2. Ⓐ ● Ⓒ Ⓓ	29. Ⓐ ● Ⓒ Ⓓ	56. Ⓐ Ⓑ Ⓒ ●
3. Ⓐ Ⓑ ● Ⓓ	30. Ⓐ Ⓑ Ⓒ ●	57. ● Ⓑ Ⓒ Ⓓ
4. Ⓐ ● Ⓒ Ⓓ	31. Ⓐ Ⓑ Ⓒ ●	58. ● Ⓑ Ⓒ Ⓓ
5. Ⓐ Ⓑ ● Ⓓ	32. Ⓐ Ⓑ Ⓒ ●	59. Ⓐ Ⓑ Ⓒ ●
6. Ⓐ ● Ⓒ Ⓓ	33. Ⓐ ● Ⓒ Ⓓ	60. ● Ⓑ Ⓒ Ⓓ
7. Ⓐ Ⓑ Ⓒ ●	34. Ⓐ Ⓑ ● Ⓓ	61. ● Ⓑ Ⓒ Ⓓ
8. ● Ⓑ Ⓒ Ⓓ	35. Ⓐ Ⓑ ● Ⓓ	62. Ⓐ Ⓑ ● Ⓓ
9. Ⓐ Ⓑ Ⓒ ●	36. Ⓐ ● Ⓒ Ⓓ	63. Ⓐ ● Ⓒ Ⓓ
10. Ⓐ ● Ⓒ Ⓓ	37. Ⓐ Ⓑ ● Ⓓ	64. Ⓐ Ⓑ ● Ⓓ
11. Ⓐ ● Ⓒ Ⓓ	38. Ⓐ Ⓑ ● Ⓓ	65. ● Ⓑ Ⓒ Ⓓ
12. Ⓐ Ⓑ ● Ⓓ	39. Ⓐ Ⓑ ● Ⓓ	66. Ⓐ Ⓑ ● Ⓓ
13. Ⓐ ● Ⓒ Ⓓ	40. Ⓐ ● Ⓒ Ⓓ	67. Ⓐ ● Ⓒ Ⓓ
14. ● Ⓑ Ⓒ Ⓓ	41. Ⓐ Ⓑ Ⓒ ●	68. Ⓐ Ⓑ ● Ⓓ
15. Ⓐ Ⓑ ● Ⓓ	42. Ⓐ Ⓑ Ⓒ ●	69. Ⓐ ● Ⓒ Ⓓ
16. Ⓐ ● Ⓒ Ⓓ	43. Ⓐ Ⓑ ● Ⓓ	70. Ⓐ Ⓑ ● Ⓓ
17. ● Ⓑ Ⓒ Ⓓ	44. ● Ⓑ Ⓒ Ⓓ	71. ● Ⓑ Ⓒ Ⓓ
18. Ⓐ ● Ⓒ Ⓓ	45. Ⓐ ● Ⓒ Ⓓ	72. Ⓐ Ⓑ ● Ⓓ
19. ● Ⓑ Ⓒ Ⓓ	46. Ⓐ ● Ⓒ Ⓓ	73. ● Ⓑ Ⓒ Ⓓ
20. ● Ⓑ Ⓒ Ⓓ	47. Ⓐ ● Ⓒ Ⓓ	74. Ⓐ Ⓑ ● Ⓓ
21. Ⓐ Ⓑ Ⓒ ●	48. Ⓐ Ⓑ Ⓒ ●	75. ● Ⓑ Ⓒ Ⓓ
22. ● Ⓑ Ⓒ Ⓓ	49. Ⓐ Ⓑ ● Ⓓ	76. Ⓐ Ⓑ Ⓒ ●
23. Ⓐ Ⓑ ● Ⓓ	50. Ⓐ Ⓑ ● Ⓓ	77. Ⓐ Ⓑ Ⓒ ●
24. Ⓐ ● Ⓒ Ⓓ	51. Ⓐ Ⓑ ● Ⓓ	78. Ⓐ Ⓑ Ⓒ ●
25. Ⓐ Ⓑ ● Ⓓ	52. Ⓐ Ⓑ ● Ⓓ	79. Ⓐ ● Ⓒ Ⓓ
26. Ⓐ Ⓑ ● Ⓓ	53. Ⓐ Ⓑ Ⓒ ●	80. Ⓐ ● Ⓒ Ⓓ
27. Ⓐ Ⓑ Ⓒ ●	54. ● Ⓑ Ⓒ Ⓓ	

Solutions

Practice Exam

1. The total dynamic head provided by the pump, h_{pump}, is that required to overcome the pressure of the pressurized tank, the friction and fitting losses, h_{losses}, and the static discharge head to the pressurized tank waterline, less the positive static head of the water column to the storage tank above.

Bernoulli's equation describes the energy (as head) relationships as

$$h_{\text{pump}} + \frac{p_1}{\gamma_1} + \frac{v_1^2}{2g} + z_1 = \frac{p_2}{\gamma_2} + \frac{v_2^2}{2g} + z_2 + h_{\text{losses}}$$

The equation can be rearranged to solve for the required pump head, choosing the water surface in the open tank as point 1 and the water surface of the pressurized tank as point 2. The velocities v_1 and v_2 are approximately zero.

$$\begin{aligned}
h_{\text{pump}} &= \frac{p_2 - p_1}{\gamma} + \frac{v_2^2 - v_1^2}{2g} + z_2 - z_1 + h_{\text{losses}} \\
&= \frac{\left(15 \frac{\text{lbf}}{\text{in}^2} - 0 \frac{\text{lbf}}{\text{in}^2}\right)\left(12 \frac{\text{in}}{\text{ft}}\right)^2}{62.4 \frac{\text{lbf}}{\text{ft}^3}} \\
&\quad + 7 \text{ ft water} - 15 \text{ ft water} + 12 \text{ ft water} \\
&= 38.62 \text{ ft water} \quad (39 \text{ ft water})
\end{aligned}$$

The answer is (C).

2. The heat added by the compressor comes from the work done by the compressor. The COP of the process is given as 4.4. The COP is the refrigeration effect divided by the power.

$$\text{COP} = \frac{\dot{q}_{\text{entering}}}{P_{\text{in}}}$$

$$P_{\text{in}} = \frac{\dot{q}_{\text{entering}}}{\text{COP}}$$

The capacity required by the evaporator to cool the water is

$$\begin{aligned}
\dot{q}_{\text{evaporator}} &= \dot{m} c_p \Delta T \\
&= \dot{V} \rho c_p (T_{\text{entering}} - T_{\text{leaving}}) \\
&= \frac{\left(\dfrac{640 \frac{\text{gal}}{\text{min}}}{7.48 \frac{\text{gal}}{\text{ft}^3}}\right)\left(60 \frac{\text{min}}{\text{hr}}\right)}{12,000 \frac{\text{Btu}}{\text{ton-hr}}} \\
&\quad \times \frac{\left(62.4 \frac{\text{lbm}}{\text{ft}^3}\right)\left(1.0 \frac{\text{Btu}}{\text{lbm-°F}}\right)}{} \\
&\quad \times (55°\text{F} - 43°\text{F}) \\
&= 320 \text{ tons}
\end{aligned}$$

The compressor power is

$$\begin{aligned}
P_{\text{in}} &= \frac{\dot{q}_{\text{evaporator}}}{\text{COP}} \\
&= \frac{320 \text{ tons}}{4.4} \\
&= 72.7 \text{ tons} \quad (73 \text{ tons})
\end{aligned}$$

The answer is (B).

3. A liquid's kinematic viscosity in Saybolt universal seconds is the time it takes for 60 cm³ of the liquid to flow through a calibrated tube at a controlled temperature. Intuitively, then, neither 0.90 SUS nor 32,000 SUS is a plausible answer choice for SAE 10W-30 oil.

For viscosities greater than about 240 SUS, one centistoke is equal to about 4.65 SUS. The kinematic viscosity in Saybolt universal seconds is

$$\begin{aligned}
\nu &= (110 \text{ cSt})\left(4.65 \frac{\text{SUS}}{\text{cSt}}\right) \\
&= 512 \text{ SUS} \quad (510 \text{ SUS})
\end{aligned}$$

The answer is (C).

4. The reheat load is the rate of sensible heat required to bring the air leaving the coil to the conditions required to supply the gallery room.

$$\dot{q}_{reheat} = \dot{m}c_p(T_{\text{leaving heating coil}} - T_{\text{entering heating coil}})$$

$$= \left(8000\ \frac{\text{ft}^3}{\text{min}}\right)\left(60\ \frac{\text{min}}{\text{hr}}\right)\left(0.075\ \frac{\text{lbm}}{\text{ft}^3}\right)$$

$$\times \left(0.24\ \frac{\text{Btu}}{\text{lbm-°F}}\right)(60°\text{F} - 53°\text{F})$$

$$= 60{,}480\ \text{Btu/hr} \quad (60{,}000\ \text{Btu/hr})$$

Alternate Solution

A psychrometric chart can also be used to determine the total energy change of the air, but the sensible heat load might have to be separated out from the total energy change if the process is anything other than pure sensible heating. From the psychrometric chart, at 53°F db and 52°F wb, the total energy content (i.e., the enthalpy) of the air leaving the cooling coil and entering the heating coil is $h_1 = 21.39$ Btu/lbm. The enthalpy of the air leaving the heating coil (and entering the museum) at 60°F db and 51°F dp is $h_2 = 23.02$ Btu/lbm. Points 1 and 2 are basically on a horizontal line, so the process is essentially pure sensible heating. The specific volume of the air is evaluated at point 2, since 8000 ft³/min is referenced to that point. From the psychrometric chart, $v_2 = 13.26$ ft³/lbm. The sensible heat load is

$$\dot{q}_{reheat} = \dot{m}\Delta h$$

$$= \frac{\dot{V}\Delta h}{v}$$

$$= \frac{\left(8000\ \frac{\text{ft}^3}{\text{min}}\right)\left(23.02\ \frac{\text{Btu}}{\text{lbm}} - 21.39\ \frac{\text{Btu}}{\text{lbm}}\right)}{13.26\ \frac{\text{ft}^3}{\text{lbm}}}$$

$$\times \left(60\ \frac{\text{min}}{\text{hr}}\right)$$

$$= 59{,}004\ \text{Btu/hr} \quad (60{,}000\ \text{Btu/hr})$$

The answer is (B).

5. For a steady, laminar, incompressible flow, the pressure drop is

$$\Delta p = \frac{128\mu\dot{V}L}{\pi d^4}$$

$$= \frac{(128)\left(26.37\times10^{-6}\ \frac{\text{lbf-sec}}{\text{ft}^2}\right)}{\pi(0.018\ \text{in})^4}$$

$$\times \left(3\times10^{-5}\ \frac{\text{ft}^3}{\text{sec}}\right)(3.4\ \text{ft})\left(12\ \frac{\text{in}}{\text{ft}}\right)^4$$

$$= 21{,}647\ \text{lbf/ft}^2 \quad (22{,}000\ \text{lbf/ft}^2)$$

The answer is (C).

6. The interest rate is unknown.

$$D_j = \frac{C - S_n}{n} = \frac{\$525{,}000 - \$0}{15} = \$35{,}000$$

$$\text{BV}_7 = C - jD_j = \$525{,}000 - (7)(\$35{,}000) = \$280{,}000$$

Gain on the sale of a depreciated asset is

$$\text{gain} = S - \text{BV}_7 = \$700{,}000 - \$280{,}000 = \$420{,}000$$

$$P = -\$525{,}000 + (\$700{,}000)(P/F, i\%, 7)$$
$$+ (\$35{,}000)(P/A, i\%, 7)(0.40)$$
$$- (\$420{,}000)(P/F, i\%, 7)(0.40)$$
$$- (\$25{,}000)(P/A, i\%, 7)(1 - 0.40)$$
$$+ (\$45{,}000)(P/A, i\%, 3)(1 - 0.40)$$
$$+ (\$75{,}000)(P/A, i\%, 7)(1 - 0.40)$$
$$- (\$75{,}000)(P/A, i\%, 3)(1 - 0.40)$$

Neglect the benefit of a favorable capital gains tax rate on the gain above \$525,000.

Simplify and combine terms.

$$P = -\$525{,}000$$
$$+ \left(\begin{array}{l}\$700{,}000\\ -(\$420{,}000)(0.40)\end{array}\right)(P/F, i\%, 7)$$
$$+ \left(\begin{array}{l}(\$35{,}000)(0.40)\\ -(\$25{,}000)(0.60)\\ +(\$75{,}000)(0.60)\end{array}\right)(P/A, i\%, 7)$$
$$+ \left(\begin{array}{l}(\$45{,}000)(0.60)\\ -(\$75{,}000)(0.60)\end{array}\right)(P/A, i\%, 3)$$
$$= -\$525{,}000 + (\$532{,}000)(P/F, i\%, 7)$$
$$+ (\$44{,}000)(P/A, i\%, 7)$$
$$+ (-\$18{,}000)(P/A, i\%, 3)$$

Try $i = 10\%$.

$$P = -\$525{,}000 + (\$532{,}000)(0.5132)$$
$$+ (\$44{,}000)(4.8684)$$
$$+ (-\$18{,}000)(2.4869)$$
$$= -\$82{,}532$$

Try $i = 5\%$.

$$P = -\$525{,}000 + (\$532{,}000)(0.7107)$$
$$+ (\$44{,}000)(5.7864)$$
$$+ (-\$18{,}000)(2.7232)$$
$$= \$58{,}676$$

Use linear interpolation.

$$\text{ROR} \approx 5\% + (10\% - 5\%)\left(\frac{\$58{,}676}{\$58{,}676 - (-\$82{,}532)}\right)$$
$$= 7.08\% \quad (7.1\%)$$

The answer is (B).

7. The Bernoulli equation can be used to solve this problem.

$$\frac{p_1}{\gamma} + \frac{\text{v}_1^2}{2g} + z_1 + h_p = \frac{p_2}{\gamma} + \frac{\text{v}_2^2}{2g} + z_2 + h_L$$

$$\frac{\left(100\ \frac{\text{lbf}}{\text{in}^2}\right)\left(12\ \frac{\text{in}}{\text{ft}}\right)^2}{62.4\ \frac{\text{lbf}}{\text{ft}^3}} + \frac{\text{v}_1^2}{(2)\left(32.2\ \frac{\text{ft}}{\text{sec}^2}\right)} + 0\ \text{ft} + h_p$$

$$= \frac{\left(500\ \frac{\text{lbf}}{\text{in}^2}\right)\left(12\ \frac{\text{in}}{\text{ft}}\right)^2}{62.4\ \frac{\text{lbf}}{\text{ft}^3}} + \frac{\text{v}_2^2}{(2)\left(32.2\ \frac{\text{ft}}{\text{sec}^2}\right)}$$

$$+ 30\ \text{ft} + 10\ \text{ft}$$

The velocities in both the suction and discharge lines can be found using

$$\dot{V} = \text{v}A$$

The flow rate in each section of the pipe is 7000 gpm. The velocity in the suction line is

$$\frac{7000\ \frac{\text{gal}}{\text{min}}}{\left(7.48\ \frac{\text{gal}}{\text{ft}^3}\right)\left(60\ \frac{\text{sec}}{\text{min}}\right)} = \text{v}_1\left(\frac{\pi}{4}\right)\left(\frac{20\ \text{in}}{12\ \frac{\text{in}}{\text{ft}}}\right)^2$$

$$\text{v}_1 = 7.15\ \text{ft/sec}$$

The velocity in the discharge line is

$$\frac{7000\ \frac{\text{gal}}{\text{min}}}{\left(7.48\ \frac{\text{gal}}{\text{ft}^3}\right)\left(60\ \frac{\text{sec}}{\text{min}}\right)} = \text{v}_2\left(\frac{\pi}{4}\right)\left(\frac{12\ \text{in}}{12\ \frac{\text{in}}{\text{ft}}}\right)^2$$

$$\text{v}_2 = 19.9\ \text{ft/sec}$$

Substitute into the Bernoulli equation and solve for the required pump head.

$$\frac{\left(100\ \frac{\text{lbf}}{\text{in}^2}\right)\left(12\ \frac{\text{in}}{\text{ft}}\right)^2}{62.4\ \frac{\text{lbf}}{\text{ft}^3}} + \frac{\left(7.15\ \frac{\text{ft}}{\text{sec}}\right)^2}{(2)\left(32.2\ \frac{\text{ft}}{\text{sec}^2}\right)}$$

$$+ 0\ \text{ft} + h_p$$

$$= \frac{\left(500\ \frac{\text{lbf}}{\text{in}^2}\right)\left(12\ \frac{\text{in}}{\text{ft}}\right)^2}{62.4\ \frac{\text{lbf}}{\text{ft}^3}} + \frac{\left(19.9\ \frac{\text{ft}}{\text{sec}}\right)^2}{(2)\left(32.2\ \frac{\text{ft}}{\text{sec}^2}\right)}$$

$$+ 30\ \text{ft} + 10\ \text{ft}$$

$$h_p = 968\ \text{ft}$$

The required power of the pump is

$$P = \dot{V}\gamma h_p$$

$$= \frac{\left(7000\ \frac{\text{gal}}{\text{min}}\right)\left(62.4\ \frac{\text{lbf}}{\text{ft}^3}\right)(968\ \text{ft})}{\left(7.48\ \frac{\text{gal}}{\text{ft}^3}\right)\left(60\ \frac{\text{sec}}{\text{min}}\right)\left(550\ \frac{\text{ft-lbf}}{\text{hp-sec}}\right)}$$

$$= 1713\ \text{hp} \quad (1700\ \text{hp})$$

The answer is (D).

8. The total resistance of the window glass includes the glass R-value plus the indoor and outdoor surface resistances.

$$R_{\text{total}} = R_{\text{outdoor air film}} + R_{\text{glass}} + R_{\text{indoor air film}}$$

The temperature drop between indoor and outdoor temperatures at a point in an assembly is proportional to the R-value at that point relative to the total R-value.

$$\frac{\Delta T_{\text{component}}}{\Delta T_{\text{total}}} = \frac{R_{\text{component}}}{R_{\text{total}}}$$

$$\Delta T_{\text{component}} = \frac{\Delta T_{\text{total}} R_{\text{component}}}{R_{\text{total}}}$$

The total resistance of the glass is the sum of the resistance of the glass and the reciprocal of the indoor and outdoor surface conductances.

$$R_{\text{total}} = \cfrac{1}{6.0 \cfrac{\text{Btu}}{\text{hr-ft}^2\text{-}°\text{F}}} + 1.5 \cfrac{\text{hr-ft}^2\text{-}°\text{F}}{\text{Btu}} + \cfrac{1}{1.46 \cfrac{\text{Btu}}{\text{hr-ft}^2\text{-}°\text{F}}}$$

$$= 2.35 \text{ hr-ft}^2\text{-}°\text{F/Btu}$$

The indoor surface temperature of the glass is calculated by subtracting the temperature drop due to the resistance of the indoor air film from room temperature.

$$\Delta T_{\text{indoor air film}} = (70°\text{F} - (-40°\text{F})) \left(\cfrac{\cfrac{1}{1.46 \cfrac{\text{Btu}}{\text{hr-ft}^2\text{-}°\text{F}}}}{2.35 \cfrac{\text{hr-ft}^2\text{-}°\text{F}}{\text{Btu}}} \right)$$

$$= 32.06°\text{F}$$
$$T_{\text{surface}} = T_{\text{room}} - \Delta T_{\text{indoor air film}}$$
$$= 70°\text{F} - 32.06°\text{F}$$
$$= 37.94°\text{F} \quad (38°\text{F})$$

To avoid condensation, the room air dew point must be less than 38°F. The corresponding relative humidity at room temperature is determined by locating 38°F on the saturation curve of the psychrometric chart and following a line of constant humidity to the right to 70°F.

This state point corresponds to approximately 31% relative humidity.

The answer is (A).

9. $T_1 = 100°\text{F} + 460° = 560°\text{R}$

n is the polytropic exponent. For a polytropic process,

$$T_2 = T_1 \left(\frac{p_2}{p_1} \right)^{\frac{n-1}{n}} = (560°\text{R}) \left(\cfrac{40 \cfrac{\text{lbf}}{\text{in}^2}}{120 \cfrac{\text{lbf}}{\text{in}^2}} \right)^{\frac{1.6-1}{1.6}}$$

$$= 370.9°\text{R}$$

The work done by the system per lbmol is given by

$$w = \frac{R(T_1 - T_2)}{n - 1} = \frac{\left(1545 \cfrac{\text{ft-lbf}}{\text{lbmol-}°\text{R}} \right)(560°\text{R} - 370.9°\text{R})}{1.6 - 1}$$

$$= 4.87 \times 10^5 \text{ ft-lbf/lbmol}$$

The molecular weight of air is 28.97 lbm/lbmol. The work per unit mass is

$$w = \cfrac{4.87 \times 10^5 \cfrac{\text{ft-lbf}}{\text{lbmol}}}{28.97 \cfrac{\text{lbm}}{\text{lbmol}}}$$

$$= 1.68 \times 10^4 \text{ ft-lbf/lbm} \quad (1.7 \times 10^4 \text{ ft-lbf/lbm})$$

The answer is (D).

10. Convert the resistivity to square inches. A circular mil is the area of a circle with a diameter of 0.001 in.

$$A_{\text{per cmil}} = \frac{\pi}{4} d^2 = \left(\frac{\pi}{4} \right)(0.001 \text{ in})^2 = 7.854 \times 10^{-7} \text{ in}^2$$

$$\rho = R_{\text{electrical}} A$$
$$= \left(10 \frac{\Omega\text{-cmil}}{\text{ft}} \right) \left(7.854 \times 10^{-7} \frac{\text{in}^2}{\text{cmil}} \right)$$
$$= 7.854 \times 10^{-6} \text{ }\Omega\text{-in}^2/\text{ft}$$

The cross-sectional area of the copper tubing is

$$A_{\text{tube}} = \frac{\pi}{4}(d_o^2 - d_i^2)$$
$$= \frac{\pi}{4}((3.5 \text{ in})^2 - (3.062 \text{ in})^2)$$
$$= 2.257 \text{ in}^2$$

The maximum resistance that will lead to a drop of no more than 0.6 V is

$$R_{\text{max}} = \frac{V_{\text{max}}}{I} = \frac{0.6 \text{ V}}{35 \text{ A}} = 0.01714 \text{ }\Omega$$

The maximum length of the circuit is

$$L_{\text{max}} = \frac{R_{\text{max}} A_{\text{tube}}}{\rho}$$
$$= \frac{(0.01714 \text{ }\Omega)(2.257 \text{ in}^2)}{7.854 \times 10^{-6} \cfrac{\Omega\text{-in}^2}{\text{ft}}}$$
$$= 4926 \text{ ft}$$

Two wires (pipes) are needed to complete the circuit, so the maximum distance between power supply and pump is

$$d_{\text{max}} = \frac{L_{\text{max}}}{2} = \frac{4926 \text{ ft}}{2} = 2463 \text{ ft} \quad (2500 \text{ ft})$$

The answer is (B).

11. Find the change in temperature. For water, $c \approx 1.0$ Btu/lbm-°F.

$$c = \frac{\Delta u}{\Delta T}$$
$$\Delta u = c\Delta T$$
$$= \left(1.0 \; \frac{\text{Btu}}{\text{lbm-}°\text{F}}\right)\Delta T$$

Multiply by the mass to find the total change in internal energy.

$$\Delta U = m\Delta u$$
$$= mc\Delta T$$
$$= (60 \text{ lbm})\left(1.0 \; \frac{\text{Btu}}{\text{lbm-}°\text{F}}\right)\Delta T$$

To find ΔU, use the first law of thermodynamics.

$$Q - W = \Delta U$$

To achieve the maximum possible change in temperature, assume the tank is perfectly insulated so that $Q = 0$. W is negative because work is being done on the system.

$$-W = \Delta U$$

The total work, W, done by the mixing motor is

$$W = Pt$$
$$= (2 \text{ hp})\left(550 \; \frac{\text{ft-lbf}}{\text{hp-sec}}\right)(15 \text{ min})\left(60 \; \frac{\text{sec}}{\text{min}}\right)$$
$$= 990{,}000 \text{ ft-lbf}$$
$$\Delta U = \frac{990{,}000 \text{ ft-lbf}}{778 \; \dfrac{\text{ft-lbf}}{\text{Btu}}}$$
$$= 1272 \text{ Btu}$$

Substitute to find ΔT.

$$\Delta U = (60 \text{ lbm})\left(1.0 \; \frac{\text{Btu}}{\text{lbm-}°\text{F}}\right)\Delta T$$
$$1272 \text{ Btu} = (60 \text{ lbm})\left(1.0 \; \frac{\text{Btu}}{\text{lbm-}°\text{F}}\right)\Delta T$$
$$\Delta T = 21.2°\text{F} \quad (21°\text{F})$$

The answer is (B).

12. Draw a diagram and a thermal circuit of the system.

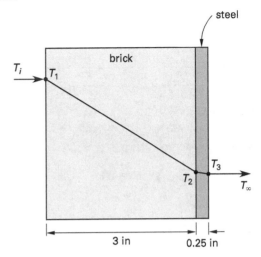

Use the given convection coefficient of $\bar{h}_{co} = 1.65$ Btu/hr-ft²-°F. The thermal resistance, R_{co}, per square foot is

$$R_{co} = \frac{1}{\bar{h}_{co}A} = \frac{1}{\left(1.65 \; \dfrac{\text{Btu}}{\text{hr-ft}^2\text{-}°\text{F}}\right)(1 \text{ ft}^2)}$$
$$= 0.606 \text{ hr-}°\text{F/Btu}$$
$$k_{br} = 0.58 \text{ Btu-ft/hr-ft}^2\text{-}°\text{F}$$
$$k_s = 26 \text{ Btu-ft/hr-ft}^2\text{-}°\text{F}$$

(The value of k_s depends on the steel composition, but it does not affect the final result.)

$$R_{br} = \frac{L_{br}}{Ak_{br}} = \frac{3 \text{ in}}{(1 \text{ ft}^2)\left(0.58 \; \dfrac{\text{Btu-ft}}{\text{hr-ft}^2\text{-}°\text{F}}\right)\left(12 \; \dfrac{\text{in}}{\text{ft}}\right)}$$
$$= 0.431 \text{ hr-}°\text{F/Btu}$$

$$R_s = \frac{L_s}{Ak_s} = \frac{0.25 \text{ in}}{(1 \text{ ft}^2)\left(26 \; \dfrac{\text{Btu-ft}}{\text{hr-ft}^2\text{-}°\text{F}}\right)\left(12 \; \dfrac{\text{in}}{\text{ft}}\right)}$$
$$= 0.0008 \text{ hr-}°\text{F/Btu}$$
$$T_1 = 1000°\text{F}$$
$$T_\infty = 70°\text{F}$$

Since T_1 and T_∞ are the only known temperatures, the heat transfer is

$$
\begin{aligned}
q &= \frac{T_1 - T_\infty}{R_{br} + R_s + R_{co}} \\
&= \frac{1000°F - 70°F}{0.431 \, \frac{hr\text{-}°F}{Btu} + 0.0008 \, \frac{hr\text{-}°F}{Btu} + 0.606 \, \frac{hr\text{-}°F}{Btu}} \\
&= 896 \text{ Btu/hr}
\end{aligned}
$$

The outside steel temperature is

$$
q = \frac{T_3 - T_\infty}{R_{co}}
$$

$$
896 \, \frac{Btu}{hr} = \frac{T_3 - 70°F}{0.606 \, \frac{hr\text{-}°F}{Btu}}
$$

$$
T_3 = 613°F \quad (610°F)
$$

The answer is (C).

13. The temperature distribution in a single-pass, parallel-flow heat exchanger is shown.

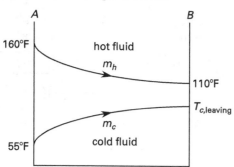

Find $T_{c,\text{leaving}}$.

$$
\begin{aligned}
q_h &= \dot{m}_h c_{p,h} \Delta T_h \\
&= \left(45{,}000 \, \frac{lbm}{hr}\right)\left(0.9 \, \frac{Btu}{lbm\text{-}°F}\right)(160°F - 110°F) \\
&= 2.025 \times 10^6 \text{ Btu/hr} \\
q_c &= \dot{m}_c c_{p,c} \Delta T_c \\
&= \left(40{,}000 \, \frac{lbm}{hr}\right)\left(1 \, \frac{Btu}{lbm\text{-}°F}\right)(T_{c,\text{leaving}} - 55°F) \\
&= \left(40{,}000 \, \frac{Btu}{hr\text{-}°F}\right)(T_{c,\text{leaving}} - 55°F)
\end{aligned}
$$

The two heat flow rates are equal, so

$$
q_h = q_c
$$

$$
2.025 \times 10^6 \, \frac{Btu}{hr} = \left(40{,}000 \, \frac{Btu}{hr\text{-}°F}\right)(T_{c,\text{leaving}} - 55°F)
$$

$$
T_{c,\text{leaving}} = 105.6°F
$$

The log mean temperature difference (LMTD) is

$$
\begin{aligned}
\text{LMTD} &= \frac{\Delta T_A - \Delta T_B}{\ln\dfrac{\Delta T_A}{\Delta T_B}} \\
&= \frac{(160°F - 55°F) - (110°F - 105.6°F)}{\ln\left(\dfrac{160°F - 55°F}{110°F - 105.6°F}\right)} \\
&= 31.7°F
\end{aligned}
$$

The heat-transfer surface area of the heat exchanger is

$$
q = UA(\text{LMTD})
$$

$$
2.025 \times 10^6 \, \frac{Btu}{hr} = \left(75 \, \frac{Btu}{hr\text{-}ft^2\text{-}°F}\right)A(31.7°F)
$$

$$
A = 851.7 \text{ ft}^2 \quad (850 \text{ ft}^2)
$$

The answer is (B).

14. Define states 1 and 2′ as shown.

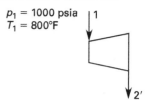

Using steam tables, $h_1 = 1388.5$ Btu/lbm, $s_1 = 1.5664$ Btu/lbm-°R, and $p_2 = 4$ psia. h_2 represents the enthalpy for a turbine that is 100% efficient. Since the turbine is isentropic, $s_1 = s_2$. Using the steam tables, find the appropriate enthalpy and entropy values at state 2′ where 2′ = 4 psia.

$$
\begin{aligned}
h_f &= 120.89 \text{ Btu/lbm} \\
s_f &= 0.21983 \text{ Btu/lbm-°R} \\
h_{fg} &= 1006.4 \text{ Btu/lbm} \\
s_{fg} &= 1.6426 \text{ Btu/lbm-°R}
\end{aligned}
$$

The steam quality at the turbine exhaust (state 2) for a 100% efficient turbine is found from the entropy relationship.

$$s = s_f + x s_{fg}$$

$$1.5664 \ \frac{\text{Btu}}{\text{lbm-}°\text{R}} = 0.21983 \ \frac{\text{Btu}}{\text{lbm-}°\text{R}}$$
$$+ x \left(1.6426 \ \frac{\text{Btu}}{\text{lbm-}°\text{R}} \right)$$

$$x = 0.82$$

The enthalpy at state 2, h_2, is

$$h_2 = h_f + x h_{fg} = 120.89 \ \frac{\text{Btu}}{\text{lbm}} + (0.82) \left(1006.4 \ \frac{\text{Btu}}{\text{lbm}} \right)$$

$$= 946.1 \ \text{Btu/lbm}$$

Since the turbine exhaust steam quality is 100%, the enthalpy at state 2′ is equal to the enthalpy of saturated vapor, h_g. From the steam tables at 4 psia,

$$h_2' = h_g = 1127.3 \ \text{Btu/lbm}$$

The efficiency of the turbine is

$$\eta_{\text{turbine}} = \frac{h_1 - h_2'}{h_1 - h_2}$$

$$= \frac{1388.5 \ \dfrac{\text{Btu}}{\text{lbm}} - 1127.3 \ \dfrac{\text{Btu}}{\text{lbm}}}{1388.5 \ \dfrac{\text{Btu}}{\text{lbm}} - 946.1 \ \dfrac{\text{Btu}}{\text{lbm}}}$$

$$= 0.59 \quad (59\%)$$

Alternate Solution

The enthalpy h_2 can be found from a Mollier diagram, thereby avoiding two calculations. To do this, first find state 1 on the Mollier diagram. It will be at the intersection of the 1000 psia pressure curve and the 800°F curve. Since a 100% efficient turbine is an isentropic process, drop straight down on the diagram until the 4 psia pressure curve is crossed. This intersection represents state 2, and the enthalpy h_2 can be read as approximately 950 Btu/lbm.

The answer is (A).

15. The general combustion reaction for a hydrocarbon, $C_n H_m$, burning with theoretical air is

$$C_n H_m + x O_2 + 3.76 x N_2 \rightarrow a CO_2 + b H_2O + 3.76 x N_2$$

$a = n$ [from carbon balance]

$2b = m$ [from hydrogen balance]

$x = n + \dfrac{m}{4}$ [from oxygen balance]

The composition of methane is CH_4, therefore $n = 1$, and $m = 4$, so $a = 1$, $b = 2$, and $x = 2$. The combustion reaction, then, for combustion of methane in theoretical air is given by

$$CH_4 + 2O_2 + (3.76)(2)N_2$$
$$\rightarrow CO_2 + 2H_2O + (3.76)(2)N_2$$

For 125% theoretical air, the reaction is

$$CH_4 + (1.25)(2)O_2 + (1.25)(7.52)N_2$$
$$\rightarrow CO_2 + 2H_2O + (0.25)(2)O_2 + (1.25)(7.52)N_2$$

The required amounts of fuel and air are

$$m_f = (1) \left(16 \ \frac{\text{lbm}}{\text{mol}} \right) = 16 \ \text{lbm/mol}$$

$$m_a = \left((1.25)(2) + (1.25)(7.52) \right) \left(28.96 \ \frac{\text{lbm}}{\text{mol}} \right)$$

$$= 344.6 \ \text{lbm/mol}$$

The air/fuel ratio is

$$R_{\text{af}} = \frac{m_a}{m_f} = \frac{344.6 \ \dfrac{\text{lbm}}{\text{mol}}}{16 \ \dfrac{\text{lbm}}{\text{mol}}} = 21.5{:}1 \quad (22{:}1)$$

The answer is (C).

16. Since the problem involves transient heat flow, determine if the lumped capacitance approximation is valid. The average temperature of the copper is

$$\overline{T} = \frac{T_s + T_l}{2} = \frac{450°\text{C} + 200°\text{C}}{2}$$

$$= 325°\text{C}$$

$$T = 325°\text{C} + 273° = 598\text{K}$$

At this temperature, the sphere's conductivity, specific heat, and density are

$$k = 379 \ \text{W/m·K}$$

$$c_p = 417 \ \text{J/kg·K}$$

$$\rho = 8933 \ \text{kg/m}^3$$

Calculate the sphere's characteristic length, L_c.

$$L_c = \frac{V}{A_s} = \frac{\frac{4}{3}\pi R^3}{4\pi R^2} = \frac{R}{3}$$

$$= \frac{100 \text{ mm}}{3}$$

$$= 33.3 \text{ mm} \quad (33.3 \times 10^{-3} \text{ m})$$

Determine the Biot number, Bi.

$$\text{Bi} = \frac{hL_c}{k}$$

$$= \frac{\left(880 \, \dfrac{\text{W}}{\text{m}^2 \cdot \text{K}}\right)(33.3 \times 10^{-3} \text{ m})}{379 \, \dfrac{\text{W}}{\text{m} \cdot \text{K}}}$$

$$= 0.0773$$

Because the Biot number is less than 0.1, the internal thermal resistance of the sphere is negligible compared to the external thermal resistance in the oil bath. Therefore, the lumped parameter method can be used.

$$T_t = T_\infty + (T_0 - T_\infty)e^{-\text{BiFo}}$$

Solve for the Fourier number, Fo, and substitute its definition.

$$\text{Fo} = \frac{-1}{\text{Bi}} \ln\left(\frac{T_t - T_\infty}{T_0 - T_\infty}\right)$$

$$\frac{kt}{\rho c_p L_c^2} = \frac{-1}{\text{Bi}} \ln\left(\frac{T_t - T_\infty}{T_0 - T_\infty}\right)$$

Solve for t.

$$t = \frac{-\rho c_p L_c^2}{k\text{Bi}} \ln\left(\frac{T_t - T_\infty}{T_0 - T_\infty}\right)$$

$$= \frac{-\left(8933 \, \dfrac{\text{kg}}{\text{m}^3}\right)\left(417 \, \dfrac{\text{J}}{\text{kg} \cdot \text{K}}\right)(3.33 \times 10^{-2} \text{ m})^2}{\left(379 \, \dfrac{\text{W}}{\text{m} \cdot \text{K}}\right)(7.73 \times 10^{-2})}$$

$$\times \ln\left(\frac{200°\text{C} - 75°\text{C}}{450°\text{C} - 75°\text{C}}\right)$$

$$= 155 \text{ s} \quad (160 \text{ s})$$

The answer is (B).

17. The useful refrigeration provided by the chiller is determined from the change in enthalpy across the evaporator.

$$\dot{q}_{\text{refrigeration}} = \dot{m}_{\text{refrigerant}}(h_{\text{leaving}} - h_{\text{entering}})$$

The enthalpy of the refrigerant entering the evaporator on the low-pressure side will be equal to that of the refrigerant leaving the condenser as a saturated liquid. From refrigerant property tables for R-134a, the enthalpy of saturated liquid at 100 psia is 37.8 Btu/lbm (38 Btu/lbm).

The enthalpy of the refrigerant entering the compressor is determined by locating the state point for saturated vapor at the low-side pressure, extended horizontally 20°F for superheat. The refrigerant temperature at a saturated vapor pressure of 33 psia is 19.8°F (20°F). Extending a horizontal line from the saturated vapor state point at 33 psia to approximately 40°F (midway between the 20°F and 60°F temperature curves) locates the superheat process line. The extension terminates at 110.2 Btu/lbm (110 Btu/lbm).

Rearranging the refrigeration equation,

$$\dot{m}_{\text{refrigerant}} = \frac{\dot{q}_{\text{refrigeration}}}{h_{\text{leaving}} - h_{\text{entering}}}$$

$$= \frac{(255 \text{ tons})\left(12{,}000 \, \dfrac{\text{Btu}}{\text{hr-ton}}\right)}{110 \, \dfrac{\text{Btu}}{\text{lbm}} - 38 \, \dfrac{\text{Btu}}{\text{lbm}}}$$

$$= 42{,}500 \text{ lbm/hr} \quad (43{,}000 \text{ lbm/hr})$$

The answer is (A).

18. The humidity ratio of the mixed air can be determined graphically on the psychrometric chart using the lever rule or from the ratio of outside air to recirculated quantities.

$$\omega_{\text{mixed}} = \omega_{\text{return}} + \left(\frac{\dot{V}_{\text{outside}}}{\dot{V}_{\text{return}} + \dot{V}_{\text{outside}}}\right)(\omega_{\text{outside}} - \omega_{\text{return}})$$

From the psychrometric chart, the humidity ratio for outside air at 90°F db and 75°F wb is 0.0153 lbm moisture/lbm dry air.

The humidity ratio for air returning from the room at 75°F db and 50% relative humidity is 0.0093 lbm moisture /lbm dry air.

$$\omega_{\text{mixed}} = \omega_{\text{return}} + \left(\frac{\dot{V}_{\text{outside}}}{\dot{V}_{\text{return}} + \dot{V}_{\text{outside}}} \right)(\omega_{\text{outside}} - \omega_{\text{return}})$$

$$= 0.0093 \; \frac{\text{lbm moisture}}{\text{lbm air}}$$

$$+ \left(\frac{2300 \; \frac{\text{ft}^3}{\text{min}}}{7000 \; \frac{\text{ft}^3}{\text{min}} + 2300 \; \frac{\text{ft}^3}{\text{min}}} \right)$$

$$\times \left(\begin{array}{c} 0.0153 \; \dfrac{\text{lbm moisture}}{\text{lbm air}} \\ -0.0093 \; \dfrac{\text{lbm moisture}}{\text{lbm air}} \end{array} \right)$$

$$= 0.0108 \; \text{lbm moisture/lbm air}$$

$$(0.011 \; \text{lbm moisture/lbm dry air})$$

The answer is (B).

19. NPSHA represents the positive head available to introduce liquid into the pump without lowering its pressure below the vapor pressure and causing cavitation. This is determined by subtracting any reductions in head from local atmospheric pressure and subtracting the friction head and vapor pressure head of the water.

$$\text{NPSHA} = h_{\text{atmospheric}} \pm h_{\text{static}} - h_{\text{vapor pressure}} - h_{\text{friction}}$$

Atmospheric pressure is known at sea level. Static and friction head are given, and the vapor pressure of the water at 80°F is determined from the steam tables as 0.50683 lbf/in². Use $\gamma = 62.4$ lbf/ft³ and $h = p/\gamma$.

$$\text{NPSHA} = h_{\text{atmospheric}} \pm h_{\text{static}} - h_{\text{vapor pressure}} - h_{\text{friction}}$$

$$= \left(14.7 \; \frac{\text{lbf}}{\text{in}^2} \right) \left(\frac{\left(12 \; \frac{\text{in}}{\text{ft}} \right)^2}{62.4 \; \frac{\text{lbf}}{\text{ft}^3}} \right) - 17 \; \text{ft}$$

$$- \left(0.50683 \; \frac{\text{lbf}}{\text{in}^2} \right) \left(\frac{\left(12 \; \frac{\text{in}}{\text{ft}} \right)^2}{62.4 \; \frac{\text{lbf}}{\text{ft}^3}} \right) - 3.0 \; \text{ft}$$

$$= 12.8 \; \text{ft} \quad (13 \; \text{ft water})$$

The answer is (A).

20. The affinity laws relate initial and final pump flow and horsepower as

$$\frac{P_{\text{final}}}{P_{\text{initial}}} = \left(\frac{Q_{\text{final}}}{Q_{\text{initial}}} \right)^3$$

The initial horsepower is

$$P_{\text{initial}} = \dot{m} \left(\frac{g}{g_c} \right) h \left(\frac{1}{\eta_{\text{pump}}} \right) = \dot{V} \rho \left(\frac{g}{g_c} \right) h \left(\frac{1}{\eta_{\text{pump}}} \right)$$

$$= \frac{\left(250 \; \frac{\text{gal}}{\text{min}} \right) \left(8.34 \; \frac{\text{lbm}}{\text{gal}} \right) \left(\dfrac{32.2 \; \frac{\text{ft}}{\text{sec}^2}}{32.2 \; \frac{\text{ft-lbm}}{\text{lbf-sec}^2}} \right)}{33,000 \; \frac{\text{ft-lbf}}{\text{hp-min}}}$$

$$\qquad \times (75 \; \text{ft}) \left(\frac{1}{0.65} \right)$$

$$= 7.29 \; \text{hp}$$

Rearranging the affinity equation, the final flow achievable when the pump is making full use of the 10 hp motor is

$$Q_{\text{final}} = Q_{\text{initial}} \sqrt[3]{\frac{P_{\text{final}}}{P_{\text{initial}}}}$$

$$= 250 \; \frac{\text{gal}}{\text{min}} \sqrt[3]{\frac{10 \; \text{hp}}{7.29 \; \text{hp}}}$$

$$= 277.8 \; \text{gpm} \quad (280 \; \text{gpm})$$

The answer is (A).

21. The casino has no walls or windows on the building exterior and the facility runs 24 hr/day. This implies that there will not be significant variation in the cooling load.

The lighting load for the space is

$$\dot{q}_{\text{lighting}} = (40,000 \; \text{ft}^2) \left(3.75 \; \frac{\text{W}}{\text{ft}^2} \right) \left(3.413 \; \frac{\text{Btu}}{\text{hr-W}} \right)$$

$$= 511,950 \; \text{Btu/hr}$$

The equipment load for the space is

$$\dot{q}_{\text{equipment}} = (80 \; \text{kW}) \left(3413 \; \frac{\text{Btu}}{\text{hr-kW}} \right) = 273,040 \; \text{Btu/hr}$$

In *ASHRAE Handbook—Fundamentals*, active people "walking around" contribute sensible and latent loads of

250 Btu/hr and 200 Btu/hr per person, respectively. The sensible and latent people loads for the space are

$$\dot{q}_{\text{people sensible}} = (40{,}000 \text{ ft}^2)\left(\frac{120 \text{ people}}{1000 \text{ ft}^2}\right)\left(250 \frac{\frac{\text{Btu}}{\text{hr}}}{\text{person}}\right)$$

$$= 1{,}200{,}000 \text{ Btu/hr}$$

$$\dot{q}_{\text{people latent}} = (40{,}000 \text{ ft}^2)\left(\frac{120 \text{ people}}{1000 \text{ ft}^2}\right)\left(200 \frac{\frac{\text{Btu}}{\text{hr}}}{\text{person}}\right)$$

$$= 960{,}000 \text{ Btu/hr}$$

The sensible load for the space is

$$\dot{q}_{\text{sensible}} = \dot{q}_{\text{lighting}} + \dot{q}_{\text{equipment}} + \dot{q}_{\text{people sensible}}$$

$$= 511{,}950 \frac{\text{Btu}}{\text{hr}} + 273{,}040 \frac{\text{Btu}}{\text{hr}}$$

$$+ 1{,}200{,}000 \frac{\text{Btu}}{\text{hr}}$$

$$= 1{,}984{,}990 \text{ Btu/hr}$$

The latent load for the space is

$$\dot{q}_{\text{latent}} = \dot{q}_{\text{steam table}} + \dot{q}_{\text{people latent}}$$

$$= 50{,}000 \frac{\text{Btu}}{\text{hr}} + 960{,}000 \frac{\text{Btu}}{\text{hr}}$$

$$= 1{,}010{,}000 \text{ Btu/hr}$$

The total load for the space is

$$\dot{q}_{\text{total}} = \dot{q}_{\text{sensible}} + \dot{q}_{\text{latent}}$$

$$= 1{,}984{,}990 \frac{\text{Btu}}{\text{hr}} + 1{,}010{,}000 \frac{\text{Btu}}{\text{hr}}$$

$$= 2{,}994{,}990 \text{ Btu/hr} \quad (3{,}000{,}000 \text{ Btu/hr})$$

The answer is (D).

22. The ADPI method relies on the selection of a diffuser whose throw (distance traveled to the point where velocity is reduced to 50 ft/min) is related to the characteristic room dimension by published ratios, which are empirically derived for 9 ft ceilings. For the case of ceiling diffusers, the maximum ADPI is achieved for all cooling load densities when

$$\frac{t_{50}}{L} = 0.8$$

For the ceiling-mounted diffuser in this problem, the characteristic dimension, L, is the distance from the diffuser to the nearest wall, which is 6.0 ft. The correct throw is

$$t_{50} = 0.8L$$

$$= (0.8)(6.0 \text{ ft})$$

$$= 4.8 \text{ ft} \quad (5.0 \text{ ft})$$

The answer is (A).

23. From the psychrometric chart, the humidity ratio for air at 70°F and 40% relative humidity is

$$\omega = 0.0062 \text{ lbm water/lbm air}$$

As the air in the car cools, the total mass of air and moisture in the car remains the same, making the process one of constant humidity ratio. Start from the initial conditions and proceed left on the chart along a constant humidity ratio line of 0.0062 lbm/lbm to 50°F. The final relative humidity is 80%.

The answer is (C).

24. According to the Darcy equation, the friction pressure loss in a conduit can be calculated as

$$h_f = f\left(\frac{L}{D_h}\right)\left(\frac{\text{v}^2}{2g}\right)$$

The hydraulic diameter, D_h, is defined as

$$D_h = \frac{4A}{P} = \frac{(4)\left(\dfrac{18 \text{ in}}{12 \frac{\text{in}}{\text{ft}}}\right)\left(\dfrac{24 \text{ in}}{12 \frac{\text{in}}{\text{ft}}}\right)}{(2)\left(\dfrac{18 \text{ in}}{12 \frac{\text{in}}{\text{ft}}}\right) + (2)\left(\dfrac{24 \text{ in}}{12 \frac{\text{in}}{\text{ft}}}\right)}$$

$$= 1.714 \text{ ft}$$

The velocity in the duct is the flow divided by the cross-sectional area.

$$\text{v} = \frac{Q}{A} = \frac{\left(4500 \frac{\text{ft}^3}{\text{min}}\right)\left(12 \frac{\text{in}}{\text{ft}}\right)^2}{(18 \text{ in})(24 \text{ in})\left(60 \frac{\text{sec}}{\text{min}}\right)}$$

$$= 25 \text{ ft/sec}$$

The Darcy equation will give results in units of feet of the fluid of interest. However, it is traditional to measure fluid head of air in terms of "inches of water," which will require additional unit conversions in terms of length and density. To convert the head from inches of air to inches of water, multiply by the ratio of the densities in these two fluids, ρ_{air}/ρ_{water}.

$$h_f = f\left(\frac{L}{D_h}\right)\left(\frac{v^2}{2g}\right)\left(\frac{\rho_{air}}{\rho_{water}}\right)$$

$$= (0.016)\left(\frac{240 \text{ ft}}{1.714 \text{ ft}}\right)\left(\frac{\left(25 \dfrac{\text{ft}}{\text{sec}}\right)^2}{(2)\left(32.2 \dfrac{\text{ft}}{\text{sec}^2}\right)}\right)$$

$$\times \left(12 \frac{\text{in}}{\text{ft}}\right)\left(\frac{0.075 \dfrac{\text{lbm air}}{\text{ft}^3}}{62.4 \dfrac{\text{lbm water}}{\text{ft}^3}}\right)$$

$$= 0.31 \text{ in water} \quad (0.3 \text{ in water})$$

The answer is (B).

25. An energy balance must be established for the crawl space. The heat gain from the heated room above through the floor to the crawl space is equal to heat loss from the crawl space to the outdoor environment. The heat loss to the environment includes heat lost through the crawl space wall and heat lost warming up infiltrating air.

$$q_{floor} = q_{wall} + q_{inf}$$
$$U_f A_f(T_r - T_c) = U_w A_w(T_c - T_o) + \dot{m}c_p(T_c - T_o)$$

The subscripts f, r, c, w, and o respectively represent floor, room, crawl space, wall, and outdoor.

To keep the pipes from freezing, the temperature of the crawl space, T_c, must remain above 32°F.

Solving for the outdoor temperature,

$$U_f A_f(T_r - T_c) = U_w A_w T_c - U_w A_w T_o$$
$$+ \dot{m}c_p T_c - \dot{m}c_p T_o$$
$$= T_c(U_w A_w + \dot{m}c_p)$$
$$- T_o(U_w A_w + \dot{m}c_p)$$
$$T_o(U_w A_w + \dot{m}c_p) = T_c(U_w A_w + \dot{m}c_p)$$
$$- U_f A_f(T_r - T_c)$$

$$T_o = \frac{T_c(U_w A_w + \dot{m}c_p) - U_f A_f(T_r - T_c)}{U_w A_w + \dot{m}c_p}$$

$$= \frac{(32°F)\left(\begin{array}{c}\left(0.15 \dfrac{\text{Btu}}{\text{hr-ft}^2\text{-}°F}\right)(320 \text{ ft}^2) \\ + \left(900 \dfrac{\text{ft}^3}{\text{hr}}\right)\left(0.075 \dfrac{\text{lbm}}{\text{ft}^3}\right) \\ \times \left(0.24 \dfrac{\text{Btu}}{\text{lbm-}°F}\right)\end{array}\right)}{\left(\begin{array}{c}\left(0.15 \dfrac{\text{Btu}}{\text{hr-ft}^2\text{-}°F}\right)(320 \text{ ft}^2) \\ + \left(900 \dfrac{\text{ft}^3}{\text{hr}}\right)\left(0.075 \dfrac{\text{lbm}}{\text{ft}^3}\right) \\ \times \left(0.24 \dfrac{\text{Btu}}{\text{lbm-}°F}\right)\end{array}\right)}$$

$$\dfrac{- \left(0.05 \dfrac{\text{Btu}}{\text{hr-ft}^2\text{-}°F}\right)(645 \text{ ft}^2)(72°F - 32°F)}{}$$

$$= 11.9°F \quad (10°F)$$

The answer is (C).

26. From saturated water tables, at 180°F,

$$h_{fg} = 989.9 \text{ Btu/lbm}$$
$$c_p = 1.00 \text{ Btu/lbm-}°F$$

The energy needed to evaporate the water is

$$E_{vapor} = m_{vapor}h_{fg}$$

An energy balance on the system gives

$$m_{vapor}h_{fg} = m_{remaining}c_p\Delta T_{remaining}$$
$$\Delta T_{remaining} = \frac{m_{vapor}h_{fg}}{m_{remaining}c_p}$$

8% of the water evaporates, so

$$\frac{m_{vapor}}{m_{remaining}} = \frac{8\%}{92\%} = 0.0870$$

So, the temperature change in the remaining water is

$$\Delta T_{\text{remaining}} = \left(\frac{m_{\text{vapor}}}{m_{\text{remaining}}}\right)\left(\frac{h_{fg}}{c_p}\right)$$

$$= (0.0870)\left(\frac{989.9 \dfrac{\text{Btu}}{\text{lbm}}}{1.00 \dfrac{\text{Btu}}{\text{lbm-°F}}}\right)$$

$$= 86.1°F$$

The remaining water has a temperature of

$$T_{\text{remaining}} = T_{\text{initial}} - \Delta T_{\text{remaining}} = 180°F - 86.1°F$$
$$= 93.9°F \quad (94°F)$$

The answer is (C).

27. The brake horsepower required by the pump is

$$P_{\text{pump}} = \dot{m}\left(\frac{g}{g_c}\right)h\left(\frac{1}{\eta_{\text{pump}}}\right)$$

$$= \dot{V}\rho\left(\frac{g}{g_c}\right)h\left(\frac{1}{\eta_{\text{pump}}}\right)$$

$$= \frac{\left(6 \dfrac{\text{gal}}{\text{min}}\right)\left(8.34 \dfrac{\text{lbm}}{\text{gal}}\right)\left(\dfrac{32.2 \dfrac{\text{ft}}{\text{sec}^2}}{32.2 \dfrac{\text{ft-lbm}}{\text{lbf-sec}^2}}\right)}{33,000 \dfrac{\text{ft-lbf}}{\text{hp-min}}}$$
$$\quad \times (140 \text{ ft})\left(\dfrac{1}{0.60}\right)$$

$$= 0.354 \text{ hp}$$

The solar array area is determined by dividing the power requirement by the product of the incident power available, I_{incident}, and the array collection efficiency, η_{array}.

$$A = \frac{P_{\text{pump}}}{\eta_{\text{array}}I_{\text{incident}}}$$

$$= \frac{(0.354 \text{ hp})\left(746 \dfrac{\text{W}}{\text{hp}}\right)}{(0.12)\left(28 \dfrac{\text{W}}{\text{ft}^2}\right)}$$

$$= 78.6 \text{ ft}^2 \quad (80 \text{ ft}^2)$$

The answer is (D).

28. The saturation efficiency for an air washer describes the completeness to which the dry-bulb temperature is reduced to the theoretical minimum wet-bulb temperature.

$$\eta_{\text{sat}} = \frac{T_{\text{db,air,entering}} - T_{\text{db,air,leaving}}}{T_{\text{db,air,entering}} - T_{\text{wb}}}$$

The leaving dry-bulb temperature for this process is

$$T_{\text{db,air,leaving}} = T_{\text{db,air,entering}}$$
$$\quad -\eta_{\text{sat}}(T_{\text{db,air,entering}} - T_{\text{wb}})$$
$$= 92°F - (0.84)(92°F - 57°F)$$
$$= 62.6°F$$

The sensible pre-cooling benefit is

$$\dot{q}_{\text{sensible}} = \dot{m}c_p(T_{\text{db,air,entering}} - T_{\text{db,air,leaving}})$$
$$= \dot{V}\rho c_p(T_{\text{db,air,entering}} - T_{\text{db,air,leaving}})$$
$$= \left(20,000 \dfrac{\text{ft}^3}{\text{min}}\right)\left(60 \dfrac{\text{min}}{\text{hr}}\right)\left(0.075 \dfrac{\text{lbm}}{\text{ft}^3}\right)$$
$$\quad \times \left(0.24 \dfrac{\text{Btu}}{\text{lbm-°F}}\right)(92°F - 62.6°F)$$
$$= 635,040 \text{ Btu/hr}$$

The chiller loads are traditionally stated in units of tons of refrigeration.

$$\dot{q}_{\text{sensible}} = \frac{635,040 \dfrac{\text{Btu}}{\text{hr}}}{12,000 \dfrac{\text{Btu}}{\text{hr-ton}}} = 52.92 \text{ tons} \quad (53 \text{ tons})$$

The answer is (C).

29. The only load is the heat loss through the brick wall. The heat loss can be calculated as

$$q_{\text{wall}} = U_{\text{wall}}A_{\text{wall}}\Delta T$$

The overall heat transfer coefficient (U-factor) is calculated by taking the reciprocal of the total thermal resistance of the wall assembly, including air surfaces. If the construction of the wall is thermally homogenous (continuous without conductive penetrations), resistance of the wall is the simple sum of the individual resistances. Individual resistances can be obtained from a wide variety of sources, the most comprehensive being the *ASHRAE Handbook—Fundamentals*. Typical resistance values for the individual wall components are listed in the following table.

wall components	resistance $\left(\dfrac{\text{hr-ft}^2\text{-}°\text{F}}{\text{Btu}}\right)$
8 in brick	0.80
2 in polystyrene	8.00
⅝ in gypsum board	0.56
outdoor air film	0.17
indoor air film (vertical)	0.68
total resistance	10.21

The wall overall heat transfer coefficient, U_{wall}, is

$$
\begin{aligned}
U_{\text{wall}} &= \frac{1}{R_{\text{wall}}} \\
&= \frac{1}{10.21 \ \frac{\text{hr-ft}^2\text{-}°\text{F}}{\text{Btu}}} \\
&= 0.098 \ \text{Btu/hr-ft}^2\text{-}°\text{F}
\end{aligned}
$$

The heat loss through the wall is

$$
\begin{aligned}
q_{\text{wall}} &= U_{\text{wall}} A_{\text{wall}} \Delta T \\
&= \left(0.098 \ \frac{\text{Btu}}{\text{hr-ft}^2\text{-}°\text{F}}\right)(800 \ \text{ft}^2)(74°\text{F} - 10°\text{F}) \\
&= 5018 \ \text{Btu/hr}
\end{aligned}
$$

The heating capacity is

$$
\begin{aligned}
q_{\text{heater}} &= \frac{\left(5018 \ \frac{\text{Btu}}{\text{hr}}\right)(1.0 + 0.25)}{3412 \ \frac{\text{Btu}}{\text{kW-hr}}} \\
&= 1.84 \ \text{kW} \quad (1.9 \ \text{kW})
\end{aligned}
$$

The answer is (B).

30. The sensible capacity of the coil can be calculated as

$$
\begin{aligned}
\dot{q}_{\text{sensible}} &= \dot{m} c_p (T_{\text{entering,db}} - T_{\text{leaving,db}}) \\
&= \dot{V} \rho c_p (T_{\text{db,air,entering}} - T_{\text{db,air,leaving}})
\end{aligned}
$$

The airflow through the coil is the product of the velocity of the air and the cross-sectional area of the coil.

$$
\begin{aligned}
\dot{V} &= \text{v} A = \left(450 \ \frac{\text{ft}}{\text{min}}\right)(3 \ \text{ft})(4 \ \text{ft}) \\
&= 5400 \ \text{ft}^3/\text{min}
\end{aligned}
$$

The sensible capacity of the coil is

$$
\begin{aligned}
\dot{q}_{\text{sensible}} &= \dot{V} \rho c_p (T_{\text{db,air,entering}} - T_{\text{db,air,leaving}}) \\
&= \left(5400 \ \frac{\text{ft}^3}{\text{min}}\right)\left(60 \ \frac{\text{min}}{\text{hr}}\right)\left(0.075 \ \frac{\text{lbm}}{\text{ft}^3}\right) \\
&\quad \times \left(0.24 \ \frac{\text{Btu}}{\text{lbm-}°\text{F}}\right)(80°\text{F} - 56°\text{F}) \\
&= 139{,}968 \ \text{Btu/hr}
\end{aligned}
$$

The grand sensible heat ratio of the coil is the ratio of the sensible capacity to the total capacity.

$$
\text{GSHR} = \frac{\dot{q}_{\text{sensible}}}{\dot{q}_{\text{total}}}
$$

The total coil capacity is determined from the definition

$$
\begin{aligned}
\dot{q}_{\text{total}} &= \frac{\dot{q}_{\text{sensible}}}{\text{GSHR}} = \frac{139{,}968 \ \frac{\text{Btu}}{\text{hr}}}{0.70} \\
&= 199{,}954 \ \text{Btu/hr} \quad (200{,}000 \ \text{Btu/hr})
\end{aligned}
$$

The answer is (D).

31. The COP is the rate of useful cooling divided by the power the compressor requires to accomplish the cooling.

$$
\text{COP} = \frac{\dot{q}_{\text{cooling}}}{P}
$$

The load rejected to the cooling tower includes both the heat rejected as cooling load and the required compressor energy. The total load rejected to the cooling tower is

$$
\dot{q}_{\text{condenser}} = \dot{m} c_p \Delta T_{\text{condenser}}
$$

$$
= \frac{\begin{pmatrix} \left(2530 \ \frac{\text{gal}}{\text{min}}\right)\left(8.34 \ \frac{\text{lbm}}{\text{gal}}\right)\left(60 \ \frac{\text{min}}{\text{hr}}\right) \\ \times \left(1.0 \ \frac{\text{Btu}}{\text{lbm-}°\text{F}}\right)(92°\text{F} - 83°\text{F}) \end{pmatrix}}{12{,}000 \ \frac{\text{Btu}}{\text{hr-ton}}}
$$

$$
= 949.5 \ \text{tons} \quad (950 \ \text{tons})
$$

The compressor heat is the difference between the total load rejected to the condenser and the cooling capacity of the chiller.

$$
\begin{aligned}
\dot{q}_{\text{compressor}} &= \dot{q}_{\text{rejected}} - \dot{q}_{\text{chiller}} \\
&= 950 \ \text{tons} - 840 \ \text{tons} \\
&= 110 \ \text{tons}
\end{aligned}
$$

The chiller COP is

$$\text{COP} = \frac{\dot{q}_{\text{chiller}}}{\dot{q}_{\text{compressor}}} = \frac{840 \text{ tons}}{110 \text{ tons}}$$
$$= 7.67 \quad (8)$$

The answer is (D).

32. Airfoil fans have the highest efficiency, due to the airfoil contour of each of their blades. The fans with the next highest efficiency are the backward-inclined and backward-curved fans. Forward-curved fans have the lowest efficiency.

The answer is (D).

33. Convert temperature from Fahrenheit to Rankine, and calculate the initial pressure in the bottle. Obtain the gas constant, R, from a table.

$$T_R = T_{\circ\text{F}} + 460^\circ = 100^\circ\text{F} + 460^\circ$$
$$= 560^\circ\text{R}$$
$$R_{\text{nitrogen}} = 55.16 \text{ ft-lbf/lbm-}^\circ\text{R}$$
$$p = \frac{m_1 R T}{V}$$

Calculate the mass remaining in the bottle.

$$m_2 = \frac{pV}{RT}$$
$$= \frac{\left(150 \dfrac{\text{lbf}}{\text{in}^2}\right)\left(12 \dfrac{\text{in}}{\text{ft}}\right)^2 (6 \text{ ft}^3)}{\left(55.16 \dfrac{\text{ft-lbf}}{\text{lbm-}^\circ\text{R}}\right)(560^\circ\text{R})}$$
$$= 4.19 \text{ lbm}$$

Calculate the mass released.

$$m_{\text{released}} = m_1 - m_2 = 5 \text{ lbm} - 4.19 \text{ lbm}$$
$$= 0.81 \text{ lbm} \quad (0.80 \text{ lbm})$$

The answer is (B).

34. The ventilation load is the rate at which sensible and latent heat must be removed from the ventilation air to reduce it to the return state.

$$\dot{q}_{\text{total}} = \dot{m}\Delta h = \dot{V}\rho(h_{\text{outdoor}} - h_{\text{room}})$$

From the psychrometric chart at 94°F db and 72°F wb,

$$h_{\text{outdoor}} = 35.6 \text{ Btu/lbm}$$

At 75°F and 45% relative humidity,

$$h_{\text{room}} = 26.8 \text{ Btu/lbm}$$
$$\dot{q}_{\text{total}} = \dot{V}\rho(h_{\text{outdoor}} - h_{\text{room}})$$
$$= \left(850 \dfrac{\text{ft}^3}{\text{min}}\right)\left(60 \dfrac{\text{min}}{\text{hr}}\right)\left(0.0733 \dfrac{\text{lbm}}{\text{ft}^3}\right)$$
$$\times \left(35.6 \dfrac{\text{Btu}}{\text{lbm}} - 26.8 \dfrac{\text{Btu}}{\text{lbm}}\right)$$
$$= 32,897 \text{ Btu/hr} \quad (30,000 \text{ Btu/hr})$$

The answer is (A).

35. The reheat coil must have sufficient heating capacity to offset winter space heat loss, as well as the colder supply air.

$$\dot{q}_{\text{coil}} = \dot{q}_{\text{space}} + \dot{q}_{\text{reheat}}$$

The reheat load is given by

$$\dot{q}_{\text{reheat}} = \dot{m} c_p (T_{\text{room}} - T_{\text{supply}})$$
$$= \dot{V}_{\text{min}} \rho_{\text{air}} c_p (T_{\text{room}} - T_{\text{supply}})$$

The lowest airflow is the product of the minimum stop fraction and the maximum flow.

$$\dot{V}_{\text{min}} = F_{\text{min}} \dot{V}_{\text{max}} = (0.30)\left(2400 \dfrac{\text{ft}^3}{\text{min}}\right)$$
$$= 720 \text{ ft}^3/\text{min}$$

The reset chart indicates that the supply air temperature will be reset to 64°F when the outdoor temperature is 10°F.

$$\dot{q}_{\text{reheat}} = \dot{V}_{\text{min}} \rho_{\text{air}} c_p (T_{\text{room}} - T_{\text{supply}})$$
$$= \left(720 \dfrac{\text{ft}^3}{\text{min}}\right)\left(60 \dfrac{\text{min}}{\text{hr}}\right)\left(0.075 \dfrac{\text{lbm}}{\text{ft}^3}\right)$$
$$\times \left(0.24 \dfrac{\text{Btu}}{\text{lbm-}^\circ\text{F}}\right)(72^\circ\text{F} - 64^\circ\text{F})$$
$$= 6220 \text{ Btu/hr}$$

The total heating coil load is

$$\dot{q}_{\text{coil}} = \dot{q}_{\text{space}} + \dot{q}_{\text{reheat}}$$
$$= 45,000 \dfrac{\text{Btu}}{\text{hr}} + 6220 \dfrac{\text{Btu}}{\text{hr}}$$
$$= 51,220 \text{ Btu/hr} \quad (51,000 \text{ Btu/hr})$$

The answer is (C).

36. The time required for the steam to cool to the ambient temperature is determined by dividing the energy reduction (from 212°F saturated steam to 60°F water) by the rate of heat loss from the pipe.

$$t = \frac{E_{\text{initial}} - E_{\text{final}}}{\dot{q}_{\text{pipe}}}$$

From standard pipe dimension schedules, the interior diameter of a 2 in schedule-40 pipe is 2.067 in. The volume of the steam in the pipe is

$$
\begin{aligned}
V &= A_i L \\
&= \frac{\pi}{4} d^2 L \\
&= \left(\frac{\pi}{4}\right)\left(\frac{2.067 \text{ in}}{12 \frac{\text{in}}{\text{ft}}}\right)^2 (80 \text{ ft}) \\
&= 1.864 \text{ ft}^3
\end{aligned}
$$

From the saturated steam tables, the specific volume of saturated 212°F steam is

$$v = 26.80 \text{ ft}^3/\text{lbm}$$

The mass of the steam in the pipe is

$$
\begin{aligned}
m_{\text{steam}} &= \rho V = \frac{V}{v} \\
&= \frac{1.864 \text{ ft}^3}{26.80 \frac{\text{ft}^3}{\text{lbm}}} \\
&= 0.0696 \text{ lbm}
\end{aligned}
$$

The initial enthalpy of the saturated 212°F steam is 1150.5 Btu/lbm.

Since the steam is initially saturated, any cooling will result in some condensation. At 60°F, all the steam will have condensed, and there will be 0.0696 lbm of 60°F liquid water in the pipe. The enthalpy of 60°F water is approximately 28.08 Btu/lbm. The total energy loss is

$$
\begin{aligned}
\Delta E_{\text{steam}} &= E_{\text{initial}} - E_{\text{final}} \\
&= m_{\text{steam}} \Delta h \\
&= (0.0696 \text{ lbm})\left(1150.5 \frac{\text{Btu}}{\text{lbm}} - 28.08 \frac{\text{Btu}}{\text{lbm}}\right) \\
&= 78.12 \text{ Btu}
\end{aligned}
$$

The time to cool is

$$
\begin{aligned}
t &= \frac{\Delta E_{\text{steam}}}{\dot{q}_{\text{pipe}}} \\
&= \frac{(78.12 \text{ Btu})\left(60 \frac{\text{min}}{\text{hr}}\right)}{6200 \frac{\text{Btu}}{\text{hr}}} \\
&= 0.756 \text{ min} \quad (1 \text{ min})
\end{aligned}
$$

The answer is (B).

37. The entropy of the steam entering the turbine at 1200°F and 700 psia is

$$s_3 = 1.7686 \text{ Btu/lbm-°R}$$

For maximum efficiency, the entropy of the steam entering the condenser, s_4, has the same value. The quality can be found from the equation

$$
\begin{aligned}
s_3 &= s_4 = s_f + x(s_g - s_f) \\
x &= \frac{s_3 - s_f}{s_g - s_f}
\end{aligned}
$$

s_f and s_g are the entropies of the saturated liquid and vapor, respectively, from a saturated steam table for a pressure of 2 psi.

$$
\begin{aligned}
x &= \frac{s_3 - s_f}{s_g - s_f} = \frac{1.7686 \frac{\text{Btu}}{\text{lbm-°R}} - 0.1750 \frac{\text{Btu}}{\text{lbm-°R}}}{1.9195 \frac{\text{Btu}}{\text{lbm-°R}} - 0.1750 \frac{\text{Btu}}{\text{lbm-°R}}} \\
&= 0.9135
\end{aligned}
$$

The enthalpy of the steam entering the condenser is

$$h_4 = h_f + x(h_g - h_f)$$

h_f and h_g are the enthalpies of the saturated liquid and vapor, respectively, and are taken from a saturated steam table for a pressure of 2 psia.

$$
\begin{aligned}
h_4 &= h_f + x(h_g - h_f) \\
&= 94.02 \frac{\text{Btu}}{\text{lbm}} + (0.9135)\left(1115.8 \frac{\text{Btu}}{\text{lbm}} - 94.02 \frac{\text{Btu}}{\text{lbm}}\right) \\
&= 1027.4 \text{ Btu/lbm}
\end{aligned}
$$

The enthalpy between the turbine and boiler can be found in a superheated steam table for 1200°F and 700 psia, and is

$$h_3 = 1625.9 \text{ Btu/lbm}$$

The enthalpy of the saturated 2 psia liquid is

$$h_1 = h_f = 94.02 \text{ Btu/lbm}$$

The efficiency of the cycle is

$$
\begin{aligned}
\eta &= \frac{W}{Q} \\
&= \frac{h_3 - h_4}{h_3 - h_1} \\
&= \frac{1625.9 \ \dfrac{\text{Btu}}{\text{lbm}} - 1027.4 \ \dfrac{\text{Btu}}{\text{lbm}}}{1625.9 \ \dfrac{\text{Btu}}{\text{lbm}} - 94.02 \ \dfrac{\text{Btu}}{\text{lbm}}} \\
&= 0.391 \quad (39\%)
\end{aligned}
$$

The answer is (C).

38. Since the temperatures of water entering and leaving the condenser are fixed, the required flow is determined by the total load to be rejected, which includes the load absorbed by the evaporator in addition to the heat added by the compressor. The relationship for the condenser water temperature rise is

$$\dot{q}_{\text{rejected}} = \dot{m}_{\text{condenser water}} c_p \Delta T$$

The ΔT term represents the water temperature change, and the load to be rejected is

$$\dot{q}_{\text{rejected}} = \dot{q}_{\text{refrigeration}} + \dot{q}_{\text{comp heat}}$$

The required condenser water can be determined as

$$\dot{m}_{\text{condenser water}} = \frac{\dot{q}_{\text{refrigeration}} + \dot{q}_{\text{comp heat}}}{c_p \Delta T}$$

The refrigeration capacity of the chiller is the refrigerant mass flow rate times the refrigeration effect.

$$
\begin{aligned}
\dot{q}_{\text{refrigeration}} &= \dot{m}_{\text{refrigerant}}(h_{\text{evap leaving}} - h_{\text{evap entering}}) \\
&= \dot{m}_{\text{refrigerant}}(h_{\text{comp entering}} - h_{\text{evap entering}})
\end{aligned}
$$

The enthalpy of the refrigerant mixture entering the chiller evaporator will be equal to that of the saturated liquid leaving the chiller condenser, since there is no change in enthalpy when the refrigerant goes through the throttling valve. Refrigerant enthalpies are found in an R-22 property table. The enthalpy of the saturated liquid at 90°F is

$$h_{\text{evap entering}} = 36.12 \text{ Btu/lbm}$$

The enthalpy of the refrigerant entering the compressor is equal to the enthalpy of the saturated vapor at 10°F, since there is no superheat. The enthalpy of the saturated vapor at 10°F is

$$h_{\text{comp entering}} = 105.27 \text{ Btu/lbm}$$

The useful refrigeration is

$$
\begin{aligned}
\dot{q}_{\text{refrigeration}} &= \dot{m}_{\text{refrigerant}}(h_{\text{comp entering}} - h_{\text{evap entering}}) \\
&= \left(117{,}000 \ \frac{\text{lbm}}{\text{hr}}\right) \\
&\quad \times \left(105.27 \ \frac{\text{Btu}}{\text{lbm}} - 36.12 \ \frac{\text{Btu}}{\text{lbm}}\right) \\
&= 8{,}090{,}433 \text{ Btu/hr}
\end{aligned}
$$

Based on the definition of COP, the compressor work can be determined.

$$
\begin{aligned}
\dot{q}_{\text{comp}} &= \frac{\dot{q}_{\text{refrigeration}}}{\text{COP}} \\
&= \frac{8{,}090{,}433 \ \dfrac{\text{Btu}}{\text{hr}}}{5.5} \\
&= 1{,}470{,}988 \text{ Btu/hr}
\end{aligned}
$$

The required rate of heat removal is

$$
\begin{aligned}
\dot{q}_{\text{rejected}} &= \dot{q}_{\text{refrigeration}} + \dot{q}_{\text{comp}} \\
&= 8{,}090{,}433 \ \frac{\text{Btu}}{\text{hr}} + 1{,}470{,}988 \ \frac{\text{Btu}}{\text{hr}} \\
&= 9{,}561{,}421 \text{ Btu/hr}
\end{aligned}
$$

The condenser water flow that is required to remove the heat is

$$
\begin{aligned}
\dot{m}_{\text{condenser}} &= \rho \dot{V}_{\text{condenser}} \\
&= \frac{\dot{q}_{\text{rejected}}}{c_p \Delta T} \\
\dot{V}_{\text{condenser}} &= \frac{\dot{q}_{\text{rejected}}}{\rho c_p \Delta T} \\
&= \frac{9{,}561{,}421 \ \dfrac{\text{Btu}}{\text{hr}}}{\left(8.34 \ \dfrac{\text{lbm}}{\text{gal}}\right)\left(1.0 \ \dfrac{\text{Btu}}{\text{lbm-}°\text{F}}\right)(95°\text{F} - 85°\text{F})} \\
&\quad \times \left(60 \ \dfrac{\text{min}}{\text{hr}}\right) \\
&= 1910.8 \text{ gpm} \quad (1900 \text{ gpm})
\end{aligned}
$$

The answer is (C).

39. ASHRAE recommends in its *HVAC Systems and Equipment* handbook that the required volume for closed tanks with air-water interfaces should be

$$V_{\text{tank}} = V_{\text{system}} \left(\frac{\left(\frac{v_2}{v_1} - 1 \right) - 3\alpha\Delta T}{\frac{p_{\text{atm}}}{p_1} - \frac{p_{\text{atm}}}{p_2}} \right)$$

The specific volume of water at 50°F is

$$v_1 = 0.01602 \text{ ft}^3/\text{lbm}$$

The specific volume of water at 105°F is

$$v_2 = 0.01615 \text{ ft}^3/\text{lbm}$$

The minimum absolute pressure, p_1, for the tank is

$$p_1 = p_{\text{gage}} + p_{\text{atm}}$$
$$= 10 \frac{\text{lbf}}{\text{in}^2} + 14.7 \frac{\text{lbf}}{\text{in}^2}$$
$$= 24.7 \text{ lbf/in}^2$$

The maximum absolute pressure, p_2, for the tank is

$$p_2 = p_{\text{gage}} + p_{\text{atm}}$$
$$= 23 \frac{\text{lbf}}{\text{in}^2} + 14.7 \frac{\text{lbf}}{\text{in}^2}$$
$$= 37.7 \text{ lbf/in}^2$$

The expansion tank size is

$$V_{\text{tank}} = V_{\text{system}} \left(\frac{\left(\frac{v_2}{v_1} - 1 \right) - 3\alpha\Delta T}{\frac{p_{\text{atm}}}{p_1} - \frac{p_{\text{atm}}}{p_2}} \right)$$

$$= (1500 \text{ gal}) \left(\frac{\left(\frac{0.01615 \frac{\text{ft}^3}{\text{lbm}}}{0.01602 \frac{\text{ft}^3}{\text{lbm}}} - 1 \right) - (3)\left(0.0000065 \frac{\text{in}}{\text{in-°F}} \right) \times (105°F - 50°F)}{\frac{14.7 \frac{\text{lbf}}{\text{in}^2}}{24.7 \frac{\text{lbf}}{\text{in}^2}} - \frac{14.7 \frac{\text{lbf}}{\text{in}^2}}{37.7 \frac{\text{lbf}}{\text{in}^2}}} \right)$$

$$= 51.47 \text{ gal} \quad (50 \text{ gal})$$

The answer is (C).

40. The temperature of the air leaving the coil is determined by subtracting the cooling coil's effect on temperature from the dry-bulb temperature of the air entering the coil.

$$T_{\text{coil leaving}} = T_{\text{coil entering}} - \Delta T_{\text{coil}}$$

The temperature of the air entering the cooling coil is simply the mixed air dry-bulb temperature, which can be determined analytically or on the psychrometric chart using a ratio of outside air to total air quantities.

$$T_{\text{coil entering}} = T_{\text{return}} + \left(\frac{\dot{V}_{\text{outside}}}{\dot{V}_{\text{return}} + \dot{V}_{\text{outside}}} \right)$$
$$\times (T_{\text{outside}} - T_{\text{return}})$$
$$= 75°F + \left(\frac{2300 \frac{\text{ft}^3}{\text{min}}}{7000 \frac{\text{ft}^3}{\text{min}} + 2300 \frac{\text{ft}^3}{\text{min}}} \right)$$
$$\times (90°F - 75°F)$$
$$= 78.7°F$$

The change in temperature across the cooling coil is determined by the sensible heat removed from the air stream. The sensible heating relationship is $\dot{q}_{\text{sensible}} = \dot{m}c_p\Delta T$.

$$\Delta T_{\text{coil}} = \frac{\dot{q}_{\text{sensible}}}{\dot{m}c_p} = \frac{\dot{q}_{\text{sensible}}}{\dot{V}\rho c_p}$$

$$= \frac{235{,}000\ \dfrac{\text{Btu}}{\text{hr}}}{\left(7000\ \dfrac{\text{ft}^3}{\text{min}} + 2300\ \dfrac{\text{ft}^3}{\text{min}}\right)\left(0.075\ \dfrac{\text{lbm}}{\text{ft}^3}\right)}$$

$$\times\left(0.24\ \dfrac{\text{Btu}}{\text{lbm}}\right)\left(60\ \dfrac{\text{min}}{\text{hr}}\right)$$

$$= 23.4°\text{F}$$

The air temperature leaving the coil is

$$T_{\text{coil leaving}} = T_{\text{coil entering}} - \Delta T_{\text{coil}} = 78.7°\text{F} - 23.4°\text{F}$$

$$= 55.3°\text{F} \quad (55°\text{F})$$

The answer is (B).

41. The process from point 3 to point 4 describes the path of air leaving the cooling coil and traveling through distribution ductwork and the spaces served. The slope of the process line is called the *sensible heat ratio* and represents the relationship of the space sensible load to the space latent load. In both cases, the loads are positive, so the process involves both sensible heating and latent heating of the supply air.

The answer is (D).

42. Note the nonstandard units used by the code. The four air quantities are calculated to determine which one will govern. The negative pressure airflow requirement is

$$\dot{V}_1 = 2610A_e\sqrt{\Delta p} = (2610)(1.5\ \text{ft}^2)\sqrt{0.05\ \text{in water}}$$

$$= 875\ \text{ft}^3/\text{min}$$

The gross floor area airflow requirement is

$$\dot{V}_2 = 0.5A_{gf} = (0.5)(1600\ \text{ft}^2) = 800\ \text{ft}^3/\text{min}$$

The temperature rise limitation airflow requirement is

$$\dot{V}_3 = \frac{\sum \dot{q}}{1.08\Delta T} = \frac{15{,}000\ \dfrac{\text{Btu}}{\text{hr}}}{(1.08)(104°\text{F} - 90°\text{F})} = 992\ \text{ft}^3/\text{min}$$

The emergency purge airflow requirement is

$$\dot{V}_4 = 100\sqrt{G} = 100\sqrt{150\ \text{lbm}}$$

$$= 1225\ \text{ft}^3/\text{min} \quad (1200\ \text{ft}^3/\text{min})$$

The emergency purge requirement sets the requirement for this room.

The answer is (D).

43. From the first law of thermodynamics,

$$\dot{Q} = \dot{m}_{\text{out}}c_p\Delta T = \dot{m}_{\text{in}}\Delta h$$

$$\dot{m}_{\text{out}} = \frac{\dot{m}_{\text{in}}\Delta h}{c_p\Delta T}$$

From saturated water tables, for a pressure of 4 psia,

$$\Delta h = h_{fg} = 1006.0\ \text{Btu/lbm}$$

For water at 65°F,

$$c_p = 0.999\ \text{Btu/lbm-°F}$$

The mass flux of the exiting water is

$$\dot{m}_{\text{out}} = \frac{\dot{m}_{\text{in}}\Delta h}{c_p\Delta T} = \frac{\left(2200\ \dfrac{\text{lbm}}{\text{hr}}\right)\left(1006.0\ \dfrac{\text{Btu}}{\text{lbm}}\right)}{\left(0.999\ \dfrac{\text{Btu}}{\text{lbm-°F}}\right)(95°\text{F} - 65°\text{F})}$$

$$= 73{,}847\ \text{lbm/hr} \quad (74{,}000\ \text{lbm/hr})$$

The answer is (C).

44. The principle is named the *Coanda effect*, after its discoverer, Henri Coanda. The lower-pressure region above an airstream traveling parallel to the ceiling causes it to adhere to the ceiling and extend the distance traveled (the throw) before reaching a defined terminal velocity.

The answer is (A).

45. The thermal energy that can be stored in the slab depends on the mass, specific heat, and the temperature difference between the slab and the room.

$$\Delta q_{\text{slab}} = mc_p\Delta T_{\text{slab}} = V\rho c_p\Delta T_{\text{slab}}$$

The temperature change of the slab is proportional to the heat lost by the slab.

$$\Delta T_{\text{slab}} = \frac{q_{\text{slab}}}{V\rho c_p}$$

Sand and gravel concrete without significant steel reinforcing that is used in construction has an approximate density of

$$\rho = 140\ \text{lbm/ft}^3$$

The concrete has an approximate specific heat of

$$c_p = 0.22 \text{ Btu/lbm-}°\text{F}$$

The rate of heat loss between the slab and the room is

$$\dot{q} = hA_{\text{slab}}(T_{\text{surface}} - T_{\text{air}})$$
$$= \left(1.4 \ \frac{\text{Btu}}{\text{hr-ft}^2\text{-}°\text{F}}\right)(20 \text{ ft})(30 \text{ ft})(82°\text{F} - 60°\text{F})$$
$$= 18{,}480 \text{ Btu/hr}$$

The slab temperature change after 1 hr will be

$$\Delta T = \frac{\Delta \dot{q}_{\text{slab}} t}{V \rho c_p}$$

$$= \frac{\left(18{,}480 \ \dfrac{\text{Btu}}{\text{hr}}\right)(1 \text{ hr})}{(20 \text{ ft})(30 \text{ ft})\left(\dfrac{6 \text{ in}}{12 \ \frac{\text{in}}{\text{ft}}}\right)\left(140 \ \dfrac{\text{lbm}}{\text{ft}^3}\right)\left(0.22 \ \dfrac{\text{Btu}}{\text{lbm-}°\text{F}}\right)}$$

$$= 2.0°\text{F} \quad (2°\text{F})$$

The answer is (B).

46. According to the ASHRAE guidelines cited, the recommended storage tank provides an hour of probable demand, appropriately factored for the facility type.

$$V_{\text{tank}} = V_{\text{probable}} F_{\text{storage factor}}$$

The probable hot-water demand is determined by adjusting the maximum demand by an industry demand factor appropriate to the facility type.

$$\dot{V}_{\text{probable}} = \dot{V}_{\text{maximum possible}} F_{\text{demand}}$$

The maximum possible demand is most easily determined in a tabular calculation format.

type of fixture	total no. of fixtures	demand per fixture (gal/hr)	total demand per fixture type (gal/hr)
lavatory	44	2	88
bathtub	22	20	440
shower	22	30	660
kitchen sink	22	10	220
dishwasher	22	15	330
		maximum possible demand	1738

The maximum probable demand is

$$\dot{V}_{\text{probable}} = \dot{V}_{\text{maximum possible}} F_{\text{demand}}$$
$$= \left(1738 \ \frac{\text{gal}}{\text{hr}}\right)(0.30)$$
$$= 521.4 \text{ gal/hr}$$

The recommended storage tank is

$$V_{\text{tank}} = V_{\text{probable}} F_{\text{storage factor}}$$
$$= \dot{V}_{\text{probable}} t F_{\text{storage factor}}$$
$$= \left(521.4 \ \frac{\text{gal}}{\text{hr}}\right)(1 \text{ hr})(1.25)$$
$$= 651.8 \text{ gal} \quad (650 \text{ gal})$$

The answer is (B).

47. From the table, the solar heat gain factor for an east-facing window in May at 0800 is

$$\text{SHGF} = 218 \text{ Btu/hr-ft}^2$$

The shading coefficient, SC, is given as 0.87, and the overall heat transfer coefficient, U, is 1.2 Btu/hr-ft²-°F. The instantaneous heat gain is given by the equation

$$\dot{q} = (\text{SC})(\text{SHGF})A + UA(T_{\text{out}} - T_{\text{in}})$$
$$= (0.87)\left(218 \ \frac{\text{Btu}}{\text{hr-ft}^2}\right)(12 \text{ ft}^2)$$
$$+ \left(1.2 \ \frac{\text{Btu}}{\text{hr-ft}^2\text{-}°\text{F}}\right)(12 \text{ ft}^2)(42°\text{F} - 75°\text{F})$$
$$= 1800.7 \text{ Btu/hr} \quad (1800 \text{ Btu/hr})$$

The answer is (B).

48. The ratio of specific heats for air is

$$k = \frac{c_p}{c_v}$$
$$= \frac{0.240 \ \dfrac{\text{Btu}}{\text{lbm-}°\text{R}}}{0.171 \ \dfrac{\text{Btu}}{\text{lbm-}°\text{R}}}$$
$$= 1.4$$

Atmospheric pressure is 14.7 psia. The initial temperature on an absolute scale is

$$T_1 = 55°\text{F} + 460° = 515°\text{R}$$

For an isentropic process, the relationship between pressure and temperature is

$$T_2 = T_1 \left(\frac{p_2}{p_1} \right)^{\frac{k-1}{k}}$$

$$= (515°\text{R}) \left(\frac{720 \frac{\text{lbf}}{\text{in}^2}}{14.7 \frac{\text{lbf}}{\text{in}^2}} \right)^{\frac{1.4-1}{1.4}}$$

$$= 1565.6°\text{R}$$

For an isentropic process, work is

$$W = -mc_v \Delta T$$

$$= -(10 \text{ lbm}) \left(0.171 \frac{\text{Btu}}{\text{lbm-°R}} \right) (1565.6°\text{R} - 515°\text{R})$$

$$= -1796 \text{ Btu} \quad (1800 \text{ Btu})$$

The answer is (D).

49. Find Δq from conduction.

$$\Delta T = 70°\text{F} - 58°\text{F} = 12°\text{F}$$

$$q_{\text{conduction}} = UA\Delta T$$

$$= \left(0.15 \frac{\text{Btu}}{\text{hr-ft}^2\text{-°F}} \right) (10{,}000 \text{ ft}^2)(12°\text{F})$$

$$+ \left(1.10 \frac{\text{Btu}}{\text{hr-ft}^2\text{-°F}} \right) (2500 \text{ ft}^2)(12°\text{F})$$

$$+ \left(0.06 \frac{\text{Btu}}{\text{hr-ft}^2\text{-°F}} \right) (25{,}000 \text{ ft}^2)(12°\text{F})$$

$$+ \left(1.6 \frac{\text{Btu}}{\text{hr-ft-°F}} \right) (720 \text{ ft})(12°\text{F})$$

$$= 82{,}824 \text{ Btu/hr}$$

Find q from infiltration. (The humidity data is not known, so Δh cannot be determined.)

$$q_{\text{infiltration}} = 0.018 Q \Delta T$$

$$= (0.018) \left(\left(0.5 \frac{\text{air change}}{\text{hr}} \right) (600{,}000 \text{ ft}^3) \right)$$

$$\times (12°\text{F})$$

$$= 64{,}800 \text{ Btu/hr}$$

The total heat loss is

$$q_{\text{total}} = q_{\text{conduction}} + q_{\text{infiltration}}$$

$$= 82{,}824 \frac{\text{Btu}}{\text{hr}} + 64{,}800 \frac{\text{Btu}}{\text{hr}}$$

$$= 147{,}624 \text{ Btu/hr}$$

The number of unoccupied hours per heating season is

$$\left(23 \frac{\text{wk}}{\text{yr}} \right) \left(\left(14 \frac{\text{hr}}{\text{day}} \right) \left(5 \frac{\text{day}}{\text{wk}} \right) + \left(24 \frac{\text{hr}}{\text{day}} \right) \left(2 \frac{\text{day}}{\text{wk}} \right) \right) = 2714 \text{ hr/yr}$$

The energy savings is

$$\frac{\left(2714 \frac{\text{hr}}{\text{yr}} \right) \left(147{,}624 \frac{\text{Btu}}{\text{hr}} \right)}{(0.77) \left(1 \times 10^5 \frac{\text{Btu}}{\text{therm}} \right)} = 5203 \text{ therms}$$

The savings is

$$\left(0.30 \frac{\$}{\text{therm}} \right) (5203 \text{ therms}) = \$1561 \quad (\$1600)$$

This assumes instantaneous heating upon start-up and ventilation during unoccupied time.

The answer is (C).

50. The operative temperature, which describes human comfort in terms of temperature, radiation, and air velocity convection, is

$$T_o = A T_{\text{air}} + (1 - A) \overline{T}_r$$

From *ASHRAE Standard 55* App. C, for airflow between 40 ft/min and 120 ft/min,

$$A = 0.6$$

The mean radiant temperature incorporates the surrounding surfaces and their respective temperatures.

$$\overline{T}_r = F_{P-1} T_1 + F_{P-2} T_2 + \cdots + F_{P-N} T_N$$

The typical mean radiant temperature is

$$\overline{T}_r = F_{\text{glass}} T_{\text{glass}} + F_{\text{walls}} T_{\text{walls}}$$

$$= (0.04)(35°\text{F}) + (1 - 0.04)(65°\text{F})$$

$$= 63.8°\text{F}$$

The operative temperature is calculated as

$$T_o = A T_{\text{air}} + (1 - A) \overline{T}_r$$

$$= (0.6)(72°\text{F}) + (1 - 0.6)(63.8°\text{F})$$

$$= 68.7°\text{F} \quad (69°\text{F})$$

The answer is (C).

51. The procedure for determining the sound pressure level, L_p, at a distance d from the sound source has been standardized by ASHRAE and codified by many local governments in their noise ordinances. The governing equation is

$$L_p = L_{\text{source}} + 10 \log Q - 20 \log d - 0.5 \quad \text{[U.S.]}$$

Units for all terms are dB. The variable Q represents directivity and has the following values.

$Q = 1$ for uniform spherical radiation with no reflecting surfaces.

$Q = 2$ for uniform hemispherical radiation with a single reflecting surface.

$Q = 4$ for uniform quadrant radiation from a point at the intersection of 2 planes.

With the chiller installed on the ground next to the building exterior wall, a directivity factor of $Q = 4$ is appropriate.

The sound pressure level at the property line is estimated to be

$$\begin{aligned} L_p &= L_{\text{source}} + 10 \log Q - 20 \log d - 0.5 \\ &= 80 \text{ dB} + 10 \log 4 - 20 \log 40 \text{ ft} - 0.5 \\ &= 80 \text{ dB} + 6.02 \text{ dB} - 32.04 \text{ dB} - 0.5 \\ &= 53.47 \text{ dB} \quad (53 \text{ dB}) \end{aligned}$$

The answer is (C).

52. The theoretical discharge from an orifice under pressure can be calculated from Toricelli's equation using the discharge coefficient (coefficient of discharge).

$$Q = C_d A_o \sqrt{2gh}$$

However, a $\frac{1}{2}$ in orifice with a coefficient of discharge of 0.75 corresponds to the standard sprinkler, and there is no need to use theoretical methods. For a standard sprinkler, the flow for a pressure, in psig in the customary U.S. system, is given as

$$Q = K\sqrt{p} = 5.6\sqrt{p}$$

The minimum pressure of 10 psig occurs at the last sprinkler head.

$$\begin{aligned} Q &= 5.6\sqrt{p_f} = 5.6\sqrt{10 \text{ psig}} \\ &= 17.7 \text{ gal/min} \end{aligned}$$

Disregarding the velocity head, the normal pressure at the next-to-last sprinkler is the sum of the normal pressure at the last sprinkler plus the friction loss between the two sprinklers. The friction loss can be found from the Hazen-Williams equation or (more typically) from a

friction loss chart. From such a chart, the loss at 17.7 gpm for a 1 in pipe is approximately 0.1 psi/ft.

The normal pressure at the next-to-last sprinkler is

$$\begin{aligned} p &= p_f + L\Delta p = 10 \text{ psig} + (10 \text{ ft})\left(0.1 \frac{\text{psi}}{\text{ft}}\right) \\ &= 11 \text{ psig} \end{aligned}$$

The discharge at the next-to-last sprinkler is

$$\begin{aligned} Q &= 5.6\sqrt{p} = 5.6\sqrt{11 \text{ psig}} \\ &= 18.6 \text{ gpm} \quad (19 \text{ gpm}) \end{aligned}$$

The answer is (C).

53. The inner diameter of the pipe in feet is

$$d = \frac{6 \text{ in}}{12 \frac{\text{in}}{\text{ft}}} = 0.5 \text{ ft}$$

The Reynolds number is

$$\begin{aligned} \text{Re} &= \frac{\text{v} d\rho}{\mu g_c} \\ &= \frac{\left(5 \frac{\text{ft}}{\text{sec}}\right)(0.5 \text{ ft})\left(62.4 \frac{\text{lbm}}{\text{ft}^3}\right)}{\left(2.359 \times 10^{-5} \frac{\text{lbf-sec}}{\text{ft}^2}\right)\left(32.2 \frac{\text{lbm-ft}}{\text{lbf-sec}^2}\right)} \\ &= 205{,}400 \end{aligned}$$

The relative roughness is

$$\frac{\epsilon}{d} = \frac{0.0008 \text{ ft}}{0.5 \text{ ft}} = 0.0016$$

From these values and the Moody diagram or a friction factor table, the Darcy friction factor is 0.024. From the Darcy equation, the head loss is

$$\begin{aligned} h_f &= \frac{fL\text{v}^2}{2dg} \\ &= \frac{(0.024)(2000 \text{ ft})\left(5 \frac{\text{ft}}{\text{sec}}\right)^2}{(2)(0.5 \text{ ft})\left(32.2 \frac{\text{ft}}{\text{sec}^2}\right)} \\ &= 37.27 \text{ ft} \quad (37 \text{ ft}) \end{aligned}$$

The answer is (D).

54. The overall heat transfer coefficient is

$$\frac{1}{U_o} = \frac{1}{h_o} + \frac{r_o}{k}\ln\frac{r_o}{r_i} + \frac{r_o}{r_i h_i}$$

$$= \frac{1}{2.0\ \dfrac{\text{Btu}}{\text{hr-ft}^2\text{-}°\text{F}}} + \left(\frac{0.75\ \text{in}}{\left(12\ \dfrac{\text{in}}{\text{ft}}\right)\left(29\ \dfrac{\text{Btu}}{\text{hr-ft-}°\text{F}}\right)}\right)$$

$$\times \ln\frac{0.75\ \text{in}}{0.685\ \text{in}}$$

$$+ \frac{0.75\ \text{in}}{(0.685\ \text{in})\left(1500\ \dfrac{\text{Btu}}{\text{hr-ft}^2\text{-}°\text{F}}\right)}$$

$$= 0.501\ \text{hr-ft}^2\text{-}°\text{F/Btu}$$

$$U_o = \frac{1}{0.501\ \dfrac{\text{hr-ft}^2\text{-}°\text{F}}{\text{Btu}}}$$

$$= 2.0\ \text{Btu/hr-ft}^2\text{-}°\text{F}$$

The saturation temperature of 1 atm steam is 212°F. The heat loss over the entire length of uninsulated pipe is

$$\dot{q} = U_o A_o \Delta T = U_o \pi D L \Delta T$$

$$= \left(2.0\ \frac{\text{Btu}}{\text{hr-ft}^2\text{-}°\text{F}}\right)\pi\left(\frac{1.5\ \text{in}}{12\ \dfrac{\text{in}}{\text{ft}}}\right)(120\ \text{ft})$$

$$\times (212°\text{F} - 60°\text{F})$$

$$= 14{,}326\ \text{Btu/hr} \quad (14{,}000\ \text{Btu/hr})$$

The answer is (A).

55. Simple payback on investment is one of the most common indexes of economic merit used when there is a need for quick decisions. The additional investment cost is divided by the annual savings attributed solely to that investment. Payback, therefore, equals incremental cost divided by annual savings.

$$t = \frac{C}{A}$$

The AFUE is a simple rating that is intended to reflect the average load-weighted efficiency of an appliance over the course of a heating season. While the prediction of energy usage can be quite complex, the estimate of savings between one scenario and another is considered to be sufficiently accurate for decision making.

The energy consumed is equal to the actual energy required (that which does not go out the furnace exhaust) divided by the average efficiency.

$$E_{\text{consumed}} = \frac{E_{\text{required}}}{\text{AFUE}}$$

The net energy required to heat the home, regardless of system losses or fuel type, is

$$E_{\text{required}} = E_{\text{consumed}}(\text{AFUE})$$

$$= \left(1500\ \frac{\text{therms}}{\text{yr}}\right)(0.78)$$

$$= 1170\ \text{therms/yr}$$

The energy consumed by the more efficient furnace is

$$E_{\text{proposed}} = \frac{E_{\text{required}}}{\text{AFUE}} = \frac{1170\ \dfrac{\text{therms}}{\text{yr}}}{0.92}$$

$$= 1272\ \text{therms/yr}$$

The annual savings is

$$E_{\text{savings}} = E_{\text{present}} - E_{\text{proposed}}$$

$$= 1500\ \frac{\text{therms}}{\text{yr}} - 1272\ \frac{\text{therms}}{\text{yr}}$$

$$= 228\ \text{therms/yr}$$

The annual cash flow reduction is

$$A = \left(228\ \frac{\text{therms}}{\text{yr}}\right)\left(\frac{\$0.85}{\text{therm}}\right)$$

$$= \$194/\text{yr}$$

The simple payback of the investment is

$$t = \frac{C}{A}$$

$$= \frac{\$800}{\dfrac{\$194}{\text{yr}}}$$

$$= 4.1\ \text{yr} \quad (4\ \text{yr})$$

The answer is (B).

56. Use the degree day method to approximate the annual energy consumption.

$$\text{annual energy} \atop \text{consumption} = \left(\frac{\frac{\text{energy}}{°F}}{\text{hr}} \right) \left(\frac{\text{degree days}}{\text{yr}} \right) \left(\frac{\text{hr}}{\text{day}} \right)$$

For the original plan, the annual energy consumption is

$$\begin{aligned} Q_{\text{yr,1}} &= \left(0.09 \; \frac{\text{Btu}}{\text{hr-ft}^2\text{-}°F} \right) \left(3500 \; \frac{°F\text{-day}}{\text{yr}} \right) \\ &\times \left(\frac{24 \; \text{hr}}{\text{day}} \right) \\ &= 75{,}600 \; \text{Btu/ft}^2\text{-yr} \end{aligned}$$

The energy cost is

$$\begin{aligned} C_1 &= \left(\frac{\$6}{1{,}000{,}000 \; \text{Btu}} \right) \left(75{,}600 \; \frac{\text{Btu}}{\text{ft}^2\text{-yr}} \right) \\ &= \$0.4536/\text{ft}^2\text{-yr} \end{aligned}$$

Since the additional insulation will reduce the U-value by one-third, it will also reduce the cost of the energy by one-third.

$$\begin{aligned} C_2 &= \tfrac{2}{3} C_1 \\ &= \left(\frac{2}{3} \right) \left(0.4536 \; \frac{\$}{\text{ft}^2\text{-yr}} \right) \\ &= \$0.3024/\text{ft}^2\text{-yr} \end{aligned}$$

The payback period is

$$\begin{aligned} \text{payback} &= \frac{\text{cost}}{\text{savings}} \\ &= \frac{0.75 \; \dfrac{\$}{\text{ft}^2}}{0.4536 \; \dfrac{\$}{\text{ft}^2\text{-yr}} - 0.3024 \; \dfrac{\$}{\text{ft}^2\text{-yr}}} \\ &= 4.960 \; \text{yr} \quad (5 \; \text{yr}) \end{aligned}$$

The answer is (D).

57. The water vapor is at the same dry-bulb temperature as the air, 80°F. From a psychrometric chart or saturated steam tables, the water vapor saturation pressure at a dry-bulb temperature of 80°F is 0.5073 psia.

The relative humidity is the ratio of the vapor pressure to the saturation pressure of water at a given temperature.

$$\phi = \frac{p_w}{p_{\text{sat}}}$$

Solving for the partial pressure of the water vapor gives

$$\begin{aligned} p_w &= \phi p_{\text{sat}} = (0.40) \left(0.5073 \; \frac{\text{lbf}}{\text{in}^2} \right) \\ &= 0.2029 \; \text{lbf/in}^2 \quad (0.20 \; \text{psia}) \end{aligned}$$

The partial pressure of the water vapor is independent of the total pressure or the pressure of the other component, air.

The answer is (A).

58. The conductive heat loss is negligible. For one-dimensional, steady-state, radial convection, the heat flux into the tube is

$$q_r = hA\Delta T = h(\pi dL)\Delta T$$

Solving for the temperature difference between the refrigerant and the ambient air gives

$$\begin{aligned} \Delta T &= \frac{\dfrac{q_r}{L}}{\pi d h} = \frac{25 \; \dfrac{\text{W}}{\text{ft}}}{\pi \left(\dfrac{0.625 \; \text{in}}{12 \; \dfrac{\text{in}}{\text{ft}}} \right) \left(5 \; \dfrac{\text{W}}{\text{ft}^2\text{-}°F} \right)} \\ &= 30.56°F \end{aligned}$$

The refrigerant is at –5°F, so the temperature of the ambient air is

$$-5°F + 30.56°F = 25.56°F \quad (26°F)$$

The answer is (A).

59. At 80°F and a relative humidity of zero, the specific humidity of the entering air is zero. From saturated air tables or the psychrometric chart, at 110°F and a relative humidity of 100%, the specific humidity of the exiting air is 0.0610 lbm/lbm.

The channel is insulated, so it can be treated as an adiabatic, steady-flow system with two inlets and one outlet. From conservation of energy, the amount of moisture picked up by the air equals the amount of moisture evaporated from the water.

The density of water is 8.345 lbm/gal. The amount of moisture picked up by the air in gallons per minute is

$$Q = (\omega_{\text{sat}} - \omega_{\text{dry}})\dot{m}$$

$$= \frac{\left(0.0610 \, \dfrac{\text{lbm}}{\text{lbm}} - 0 \, \dfrac{\text{lbm}}{\text{lbm}}\right)}{8.345 \, \dfrac{\text{lbm}}{\text{gal}}} \times \left(\left(30 \, \dfrac{\text{lbm}}{\text{sec}}\right)\left(60 \, \dfrac{\text{sec}}{\text{min}}\right)\right)$$

$$= 13.16 \, \text{gal/min} \quad (13 \, \text{gpm})$$

The answer is (D).

60. From a table of pipe dimensions (such as MERM App. "Dimensions of Welded and Seamless Pipe"), the cross-sectional area of schedule-40 pipe with a nominal diameter of $3/4$ in is 0.0037 ft^2. The flow rate in cubic feet per second is

$$\dot{V} = \frac{48 \, \dfrac{\text{gal}}{\text{min}}}{449 \, \dfrac{\text{sec-gal}}{\text{ft}^3\text{-min}}} = 0.1069 \, \text{ft}^3/\text{sec}$$

The fluid velocity through the pipe is

$$\text{v} = \frac{\dot{V}}{A} = \frac{0.1069 \, \dfrac{\text{ft}^3}{\text{sec}}}{0.0037 \, \text{ft}^2} = 28.89 \, \text{ft/sec}$$

The total minor losses are the sum of the loss contributions of each of the fittings. There are two 90° elbows and one 45° elbow, so

$$h_m = \sum h_{L,\text{fitting}}$$

$$= \sum \left(\frac{\text{v}^2}{2g}\right)k_{\text{fitting}}$$

$$= \frac{\text{v}^2}{2g} \sum k_{\text{fitting}}$$

$$= \left(\frac{\left(28.89 \, \dfrac{\text{ft}}{\text{sec}}\right)^2}{(2)\left(32.2 \, \dfrac{\text{ft}}{\text{sec}^2}\right)}\right)(0.9 + 0.9 + 0.42)$$

$$= 28.77 \, \text{ft} \quad (29 \, \text{ft})$$

The answer is (A).

61. From the psychrometric chart, air at 40°F and 20% relative humidity has a specific humidity of 0.001 lbm/lbm. Relevant information from the chart is shown.

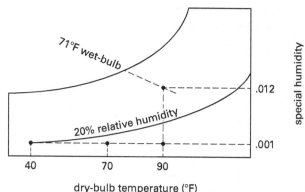

The addition of sensible heat to 70°F does not affect the specific humidity. From the psychrometric chart, the air at the final condition of 90°F dry-bulb, 71°F wet-bulb has a specific humidity of 0.012 lbm/lbm. The mass flow rate at which hot steam is needed is

$$\dot{m}_{\text{steam}} = \dot{V}_{\text{air}}\left(\frac{\omega_2 - \omega_1}{v}\right)$$

$$= \left(6000 \, \frac{\text{ft}^3}{\text{min}}\right)\left(\frac{0.012 \, \dfrac{\text{lbm}}{\text{lbm}} - 0.001 \, \dfrac{\text{lbm}}{\text{lbm}}}{12.8 \, \dfrac{\text{ft}^3}{\text{lbm}}}\right)$$

$$= 5.156 \, \text{lbm/min}$$

Converting this to volumetric flow rate using the density of water, 8.345 lbm/gal, gives

$$\dot{V} = \frac{\dot{m}_{\text{steam}}}{\rho_{\text{water}}} = \frac{5.156 \, \dfrac{\text{lbm}}{\text{min}}}{8.345 \, \dfrac{\text{lbm}}{\text{gal}}}$$

$$= 0.6179 \, \text{gal/min} \quad (0.62 \, \text{gpm})$$

The answer is (A).

62. The affinity laws indicate that when flow is reduced by 50%, speed is reduced by 50%, which in this case would be 1725 rpm. However, the pump is part of a larger application, and the system curve for this application does not start at zero head.

One-half the design flow is 500 gpm/2 = 250 gpm. According to the graph, 250 gpm occurs with a speed nearest to 2587 rpm (2600 rpm).

The answer is (C).

63. Thermal energy storage is a method of shifting some or all of a building's cooling to off-peak, nighttime hours. During off-peak hours, ice is made and stored inside one or more energy storage tanks. The stored energy is then used to cool the building the next day.

Thermal energy storage is most effective when storage can be done in off-peak hours and used later at peak hours. Thermal energy storage is least effective for an application that requires the system to operate 24 hours a day, such as a large refrigerated warehouse.

The answer is (B).

64. For a polytropic process,

$$\frac{p_2}{p_1} = \left(\frac{T_2}{T_1}\right)^{n/(n-1)}$$

n is the polytropic index. For this compression process,

$$\frac{260 \frac{\text{lbf}}{\text{in}^2}}{92 \frac{\text{lbf}}{\text{in}^2}} = \left(\frac{145°\text{F} + 460°}{25°\text{F} + 460°}\right)^{n/(n-1)}$$

$$2.82 = (1.25)^{n/(n-1)}$$

$$\ln 2.82 = \ln(1.25)^{n/(n-1)}$$

$$= \frac{n}{n-1}\ln 1.25$$

$$\frac{n}{n-1} = \frac{\ln 2.82}{\ln 1.25} = 4.64$$

$$n = \frac{4.64}{4.64 - 1} = 1.279 \quad (1.3)$$

The answer is (C).

65. To compare two alternatives using payback analysis, first determine the annual energy cost for each option.

$$C_{\text{annual}} = E_{\text{kW-hr/ton}} C_{\$/\text{kW-hr}} n_{\text{hr/yr}}$$

$$C_{\text{annual},1} = \left(0.8 \frac{\text{kW}}{\text{ton}}\right)\left(0.10 \frac{\$}{\text{kW-hr}}\right)\left(2000 \frac{\text{hr}}{\text{yr}}\right)$$

$$= \$160/\text{ton-yr}$$

$$C_{\text{annual},2} = \left(0.7 \frac{\text{kW}}{\text{ton}}\right)\left(0.10 \frac{\$}{\text{kW-hr}}\right)\left(2000 \frac{\text{hr}}{\text{yr}}\right)$$

$$= \$140/\text{ton-yr}$$

Option 2 has the higher cost efficiency. The payback period is given by the difference in installed costs divided by the difference in annual costs.

$$\text{payback period} = \frac{\Delta C_{\text{install}}}{\Delta C_{\text{annual}}}$$

$$= \frac{C_{\text{install},1} - C_{\text{install},2}}{C_{\text{annual},1} - C_{\text{annual},2}}$$

$$= \frac{350 \frac{\$}{\text{ton}} - 300 \frac{\$}{\text{ton}}}{160 \frac{\$}{\text{ton-yr}} - 140 \frac{\$}{\text{ton-yr}}}$$

$$= 2.5 \text{ yr}$$

The answer is (A).

66. Heat is transferred from the exhaust air to the makeup air. From the psychrometric chart, the specific enthalpy of the makeup air before conditioning is 9.3 Btu/lbm. Revelant information from the chart is shown.

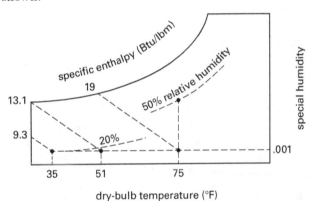

dry-bulb temperature (°F)

The exhaust air undergoes a latent heat change as it transfers heat to the makeup air. After conditioning, the exhaust air has the same humidity ratio as the makeup air. On the psychrometric chart, find the states of the exhaust air and makeup air before conditioning. Draw a horizontal line to the right from the makeup air, and draw a vertical line down from the exhaust air. The point where these two lines intersect is the state of the exhaust air after conditioning. This point has a specific enthalpy of 19 Btu/lbm. The sensible heat difference between the exhaust air after conditioning and the makeup air before conditioning is

$$19 \frac{\text{Btu}}{\text{lbm}} - 9.3 \frac{\text{Btu}}{\text{lbm}} = 9.7 \text{ Btu/lbm}$$

The sensible heat recovered is 40% of this, or

$$(0.4)\left(9.7 \frac{\text{Btu}}{\text{lbm}}\right) = 3.88 \text{ Btu/lbm} \quad (4 \text{ Btu/lbm})$$

The problem can also be solved more quickly by recognizing that, because the exhaust and makeup airflow rates are equal, a 40% sensible heat transfer results in a 40% dry-bulb temperature rise. The dry-bulb temperature of the makeup air rises from 35°F to

$$0.4 = \frac{(T_{\text{makeup}} - 35°F)}{(75°F - 35°F)}$$

$$T_{\text{makeup}} = 51°F$$

From the psychrometric chart, the conditioned makeup air at 51°F has a specific enthalpy of 13.1 Btu/lbm. The increase in specific enthalpy is

$$13.1 \frac{\text{Btu}}{\text{lbm}} - 9.3 \frac{\text{Btu}}{\text{lbm}} = 3.8 \text{ Btu/lbm} \quad (4 \text{ Btu/lbm})$$

The answer is (B).

67. The three-way valve shown is a mixing valve, which is commonly used to mix heating water to a desired temperature.

The answer is (C).

68. In variable refrigerant charge systems (varying based on load demand), the receiver, sometimes called the receiver dryer, is a component in the liquid line between the condenser and the thermostatic expansion valve. Its sole function is to store refrigerant.

An accumulator protects the compressor from liquid slugging, and it is installed on the suction side of the compressor (i.e., between the evaporator and the compressor). An expansion tank is a water system component, not a refrigerant component. In a small system, a condenser can store some refrigerant, but its primary function is to condense the refrigerant vapor.

The answer is (B).

69. *Legionella* is a bacteria found naturally in freshwater such as lakes and streams. It can become a health concern when it grows and spreads in human-made water systems such as showers and faucets, cooling towers, hot tubs, and so on.

Once *Legionella* has grown and multiplied in a building's water system, the contaminated water may spread through the air in the form of droplets small enough for people to breathe in. People who breathe in the bacteria can get Legionnaires' disease.

The most important factor in controlling *Legionella* in a building is proper maintenance of the water systems. This includes, but is not limited to, cleaning components such as cooling towers regularly, treating water systems, and testing for contamination.

Proper maintenance of water system controls and of refrigeration systems (particularly condensate water) can also help, as can changing air filters to trap the bacteria, but these factors have much less of an effect.

The answer is (B).

70. The only reason for pressure drop in this ductwork is friction. The pressure available to overcome friction will determine how fast the air will flow. For both the straight branch and the take-off branch, this pressure is equal to the pressure at the branch minus what is needed to drive the diffusers. This available pressure is

$$p_{\text{avail}} = 0.10 \text{ in wg} - 0.03 \text{ in wg}$$
$$= 0.07 \text{ in wg}$$

Convert the available pressure to inches of water gage per 100 ft of duct. For the 20 ft long straight branch,

$$\left(\frac{0.07 \text{ in wg}}{20 \text{ ft}} \right)(100 \text{ ft}) = 0.35 \text{ in wg}$$

From MERM Fig. "Standard Friction Loss in Standard Duct," using this pressure per 100 ft of duct and an air quantity of 1000 ft³/min, the needed duct diameter is nearly 11 in.

For the 30 ft long take-off branch,

$$\left(\frac{0.07 \text{ in wg}}{30 \text{ ft}} \right)(100 \text{ ft}) = 0.23 \text{ in wg}$$

From the same MERM figure, using this pressure per 100 ft of duct and an air quantity of 1000 ft³/min, the needed duct diameter is nearly 12 in.

The answer is (C).

71. The refrigerating effect is the enthalpy difference between the condition of the refrigerant when entering the evaporator and its condition when leaving. Use the pressure-enthalpy diagram in MERM App. "Pressure-Enthalpy Diagram for Refrigerant HFC-134a." When the refrigerant enters the evaporator at 34.7 psia and 15% quality, its enthalpy, h, is

$$h_{\text{in}} = 30 \text{ Btu/lbm}$$

When the refrigerant leaves the evaporator at 29.7 psia and 100% quality, its enthalpy is

$$h_{\text{out}} = 105 \text{ Btu/lbm}$$

The refrigerating effect is $h_{out} - h_{in}$, and the refrigerating capacity of the system is

$$\text{capacity} = (h_{out} - h_{in})\dot{m}$$

$$= \frac{\left(105 \, \frac{\text{Btu}}{\text{lbm}} - 30 \, \frac{\text{Btu}}{\text{lbm}}\right)}{12{,}000 \, \frac{\text{Btu}}{\text{hr-ton}}} \times \left(\left(10 \, \frac{\text{lbm}}{\text{min}}\right)\left(60 \, \frac{\text{min}}{\text{hr}}\right)\right)$$

$$= 3.75 \text{ tons} \quad (4 \text{ tons})$$

The answer is (A).

72. Waste heat refers to energy that is generated but not put to practical use. While some waste heat losses from industrial processes are inevitable, losses can be reduced by improving equipment efficiency and by installing waste heat recovery technologies. Waste heat recovery entails capturing and reusing the waste heat from industrial processes or the heat lost from air conditioning systems.

Waste heat can take the forms of both sensible and latent heat. Latent heat is often the greater part, as in combustion gas exhaust.

However, recovering latent waste heat is not as simple as recovering sensible waste heat (high temperature streams). To recover latent waste heat, the transfer material must come in direct contact with the exhausting stream and be able to absorb and evaporate liquids.

Heat recovery wheels, or thermal wheels, come in direct contact with the exhausting stream, and some have the capability to absorb moisture. The evaporation in the heat recovery wheel is a liquid-to-vapor phase change at a constant temperature. A phase change at constant temperature is a latent heat change.

Run-around, heat pipe, and heat pump are indirect methods.

The answer is (C).

73. To find the energy efficiency ratio (EER), take the cooling output in British thermal units per hour and divide it by the electrical load in watts.

$$\text{output} = (40 \text{ tons})\left(12{,}000 \, \frac{\text{Btu}}{\text{hr-ton}}\right)$$

$$= 480{,}000 \text{ Btu/hr}$$

$$\text{EER} = \frac{\text{output in Btu/hr}}{\text{load in watts}}$$

$$= \frac{480{,}000 \, \frac{\text{Btu}}{\text{hr}}}{55{,}000 \text{ W}}$$

$$= 8.727 \quad (8.7)$$

The answer is (A).

74. Use the pressure-enthalpy (p-h) diagram for HFC-134a to determine the mass flow rate and the added head. The illustration shows the p-h diagram for this cycle, including the 20°F superheat and 10°F subcooling.

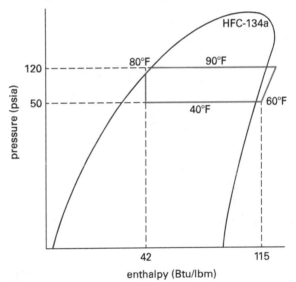

Conditions at the evaporator are

$$h_{in} = 42 \text{ Btu/lbm}$$
$$h_{out} = 115 \text{ Btu/lbm}$$

The refrigerating effect, RE, is

$$\text{RE} = h_{out} - h_{in}$$

$$= 115 \, \frac{\text{Btu}}{\text{lbm}} - 42 \, \frac{\text{Btu}}{\text{lbm}}$$

$$= 73 \text{ Btu/lbm}$$

The capacity in British thermal units per hour is

$$\dot{Q} = (3 \text{ tons})\left(12{,}000 \ \frac{\text{Btu}}{\text{hr}}\right)$$
$$= 36{,}000 \ \text{Btu/hr}$$

The mass flow rate rate, $\dot{m}$, is

$$\dot{m} = \frac{\dot{Q}}{\text{RE}}$$
$$= \frac{\left(\dfrac{36{,}000 \ \frac{\text{Btu}}{\text{hr}}}{3600 \ \frac{\text{sec}}{\text{hr}}}\right)}{73 \ \frac{\text{Btu}}{\text{lbm}}}$$
$$= 0.1370 \ \text{lbm/sec}$$

For the compressor, the head added is

$$h_A = \frac{\Delta p}{\rho} \times \frac{g_c}{g}$$
$$= \Delta p v \times \frac{g_c}{g}$$
$$= \left(\left(120 \ \frac{\text{lbf}}{\text{in}^2} - 50 \ \frac{\text{lbf}}{\text{in}^2}\right)\left(12 \ \frac{\text{in}}{\text{ft}}\right)^2\left(1 \ \frac{\text{ft}^3}{\text{lbm}}\right)\right.$$
$$\left. \times \left(\frac{32.2 \ \frac{\text{lbm-ft}}{\text{lbf-sec}^2}}{32.2 \ \frac{\text{ft}}{\text{sec}^2}}\right)\right)$$
$$= 10{,}080 \ \text{ft}$$

The horsepower needed to add 10,080 ft of head is

$$P = \dot{m} h_A \times \frac{g}{g_c}$$
$$= \left(\frac{\left(0.1370 \ \frac{\text{lbm}}{\text{sec}}\right)(10{,}080 \ \text{ft})}{550 \ \frac{\text{ft-lbf}}{\text{hp-sec}}}\right)\left(\frac{32.2 \ \frac{\text{ft}}{\text{sec}^2}}{32.2 \ \frac{\text{lbm-ft}}{\text{lbf-sec}^2}}\right)$$
$$= 2.5 \ \text{hp}$$

The answer is (C).

75. Equal friction, constant volume, and static regain are all methods of sizing air conditioning ductwork and are described in the *ASHRAE Handbook—Fundamentals*. The number of transfer units (NTU) is a method used to size cooling towers.

The answer is (A).

76. Plotting the supply air condition, the outside air condition, and the indoor design condition on a psychrometric chart, the humidity ratio for the supply air is 0.0115 lbm/lbm, the humidity ratio for the outside air is 0.00085 lbm/lbm, and the humidity ratio for the indoor design conditions is 0.0084 lbm/lbm.

Find the rate of moisture that is needed by the building, based on supply air conditions with an airflow rate of 20,000 cfm. Use an approximate value of 13.4 ft^3/lbm for the specific volume of the air.

The total rate of moisture added by the supply air is

$$W = \frac{\dot{V}(\omega_{\text{SA}} - \omega_{\text{IDC}})}{v_{\text{air}}}$$
$$= \frac{\left(\left(20{,}000 \ \frac{\text{ft}^3}{\text{min}}\right)\left(60 \ \frac{\text{min}}{\text{hr}}\right)\right)}{13.4 \ \frac{\text{ft}^3}{\text{lbm}}}$$
$$\times \left(0.0115 \ \frac{\text{lbm}}{\text{lbm}} - 0.0084 \ \frac{\text{lbm}}{\text{lbm}}\right)$$
$$= 277.6 \ \text{lbm/hr}$$

The outside air will supply only 4000 cfm. Find the needed increase in specific humidity to achieve the same humidification for the whole building.

$$W = \frac{\dot{V}(\omega_{\text{DOAS}} - \omega_{\text{OA}})}{v_{\text{air}}}$$
$$\omega_{\text{DOAS}} = \frac{W v_{\text{air}}}{\dot{V}} + \omega_{\text{OA}}$$
$$= \frac{\left(\dfrac{277.6 \ \frac{\text{lbm}}{\text{hr}}}{60 \ \frac{\text{min}}{\text{hr}}}\right)\left(13.4 \ \frac{\text{ft}^3}{\text{lbm}}\right)}{4000 \ \frac{\text{ft}^3}{\text{min}}} + 0.00085 \ \frac{\text{lbm}}{\text{lbm}}$$
$$= 0.01635 \ \text{lbm/lbm}$$

From the psychrometric chart, for a temperature of 75°F and a humidity ratio of about 0.016 lbm/lbm, the relative humidity needed for the supply air for the dedicated outdoor air system is 85%.

The answer is (D).

77. An air-side economizer can save energy by using outside air to cool the building instead of recirculated air. The load on the air handler unit (AHU) depends on the difference in enthalpies between the supply air and the return air. For this reason, the economizer should be turned on when the enthalpy of the outside air is less than the enthalpy of the return air, which is the enthalpy of the room design condition. (The refrigerating coil temperature may have to be adjusted for a new coil sensible heat factor.)

The answer is (D).

78. ASHRAE 90 is an energy efficiency standard that provides minimum requirements for the energy-efficient design of buildings except low-rise residential buildings. It also provides criteria for determining compliance with these requirements.

ASHRAE 15 is a safety code for mechanical refrigeration. ASHRAE 55 is a standard for human comfort conditions in a building. ASHRAE 62 is a standard for ventilation and outdoor air requirements for dilution of pollutants in a building.

The answer is (D).

79. The decibel (dB) is a unit that measures the effect of sound intensity on the human ear. A sound of minimal audibility is assigned the value of 0 dB.

The sound power ratio is $10^{\text{dB level}/10}$ to 1. For example, a 0 dB sound has a sound power ratio of $10^{0/10}$ to 1, or 1 to 1. A 20 dB sound has a sound power ratio of $10^{20/10}$ to 1, or 100 to 1. This means that a 20 dB sound is 100 times more powerful than a 0 dB sound.

Sound intensity is distributed spherically, so it decreases rapidly as the distance from the source increases. The rule of acoustics is to deduct 6 dB for every doubling of the distance from the source. In this case, the initial distance from the source is 10 ft, so deduct 6 dB for moving to 20 ft, another 6 dB for moving to 40 ft, and another 6 dB for moving to 80 ft.

Therefore, the sound level at 80 ft is

$$\begin{aligned} L_{80\,\text{ft}} &= L_{10\,\text{ft}} - (3)(6\ \text{dB}) \\ &= 100\ \text{dB} - 18\ \text{dB} \\ &= 82\ \text{dB} \end{aligned}$$

The answer is (B).

80. ASHRAE Standard 62.1 Table 6-1 requires 5 ft^3/min of outdoor air per person for an auditorium seating area, plus 0.06 ft^3/min per square foot based on the area of the auditorium.

The maximum estimated occupancy, according to the same table, is 150 people per 1000 ft^2, so for a 5000 ft^2 auditorium, the maximum estimated occupancy is 5×150 people $= 750$ people.

The rate of ventilation required is

$$\begin{aligned} \dot{V} &= (750\ \text{persons})\left(\frac{5\ \text{ft}^3}{\text{min-person}}\right) \\ &\quad + \left(0.06\ \frac{\text{ft}^3}{\text{min-ft}^2}\right)(5000\ \text{ft}^2) \\ &= 4050\ \text{ft}^3/\text{min} \quad (4000\ \text{ft}^3/\text{min}) \end{aligned}$$

The answer is (B).